NEW ZEALAND DEPARTMENT
OF SCIENTIFIC AND INDUSTRIAL RESEARCH

DSIR Information Series No. 150

Illustrations of New Zealand Fossils

A New Zealand Geological Survey Handbook

Compiled by I. G. Speden and I. W. Keyes

from Contributions by Staff of
Macropaleontology Section
New Zealand Geological Survey

R. P. Suggate
Director

1981
Wellington

Edited by Anne C. Wright, Science Information Division, DSIR

Cataloguing in Publication

SPEDEN, I. G.
Illustrations of New Zealand fossils: a New Zealand Geological Survey handbook/compiled by I. G. Speden and I. W. Keyes.–Wellington: Science Information Division, 1981.
(DSIR information series, ISSN 0077–9636; No. 150).
ISBN 0–477–06671–2

Front cover *Pecten novaezelandiae aotea* Fleming. Holotype of left (flat) and right valves, from Landguard Bluff, Wanganui. Castlecliffian Stage (about 500 000 years before present).

Contents

Figures

Plates

Dedication

This volume is dedicated to Sir Charles Fleming, KBE, FRS, FRSNZ, OBE, DSc (Chief Paleontologist, New Zealand Geological Survey, 1953–1977) in recognition of his numerous and notable contributions to paleontology, biology, and biogeography. It also represents an appreciation from his colleagues in Macropaleontology Section, New Zealand Geological Survey, namely A. G. Beu, R. A. Cooper, I. W. Keyes, E. McGregor, P. A. Maxwell, J. E. Simes, I. G. Speden, and G. R. Stevens.

Introduction

As stated in the dedication this "field guide" is an appreciation of the work, scientific and public, of Sir Charles Fleming. It has a second function, one that has been foremost in the thoughts of Sir Charles. This is to make readily available to the public, at both an amateur and a professional level, information which permits an understanding and enjoyment of fossils and their history.

The book is a guide to the identification of the fossils of the major time intervals currently recognised in the marine sedimentary rocks of New Zealand. Most of the fossils illustrated are marine macrofossils, that is, those that can be seen by the human eye. The majority comprise bivalves (clams), gastropods (snails), ammonites, and brachiopods (lamp shells). Some non-marine plants (Plates 15, 16), marine vertebrates (penguin bones, whale tooth; Plate 26), and a few microfossils (Hornibrook 1968), mainly conodonts (Plate 1), are illustrated.

To aid determination of the age of the rock from which the fossil was collected several figures (Fig. 4–6) are included which show the present time subdivisions of New Zealand rocks, from the oldest (Cambrian, starting 570 million years ago) to the present day. Also included are maps showing the distribution of major age classes of rocks (Fig. 1 and 2) in New Zealand. It should be remembered that hard-shelled fossils are known from only a small part, approximately one-ninth, of the time span since the earth was formed (Fig. 3).

The representative fossils illustrated are found in many parts of New Zealand, where they are present in rocks of the appropriate age. However, for many reasons the rocks of a particular region may be devoid of fossils. Common factors influencing the absence of fossils are the lack of rock outcrops, weathering of rock, unfavourable environmental conditions prevailing at the time the fossil existed, or subsequent metamorphism and deformation. Thus, over a large area along the main ranges of New Zealand rocks of Permian to Early Cretaceous age contain very few fossils (Fig. 1 and 2). Generally, the older the rocks the more difficult it becomes to collect fossils. Nevertheless, attractive fossils may be found in rocks of any age.

The fossils illustrated are selected specimens, either the original shell, external impression or internal mould, or prepared casts of these. However, the reader should note that because of poor preservation, distortion or incompleteness, or morphological variation, fossils found in rocks in the field may look different or be hard to identify. Discovery, collection, and identification of fossils are skills which require training, knowledge, and experience. Further information on the classification and occurrence of New Zealand fossils is available in scientific publications, including those listed under "Further Reading".

The plates of fossils included here were taken from the *Geology of New Zealand* (Suggate *et al.* 1979), which is the primary source for further information on the fossils and the rocks in which they are found. We are especially grateful to the Government Printer and the Department of Scientific and Industrial Research for permission to republish the plates. Professors J. D. Campbell, University of Otago, Dunedin, and J. B. Waterhouse, University of Queensland, Brisbane, kindly provided information. D. A. Francis and David Zwartz advised on format and content, Miss J. O. Palmer draughted the figures, and Mrs Fay Tonks typed the manuscript.

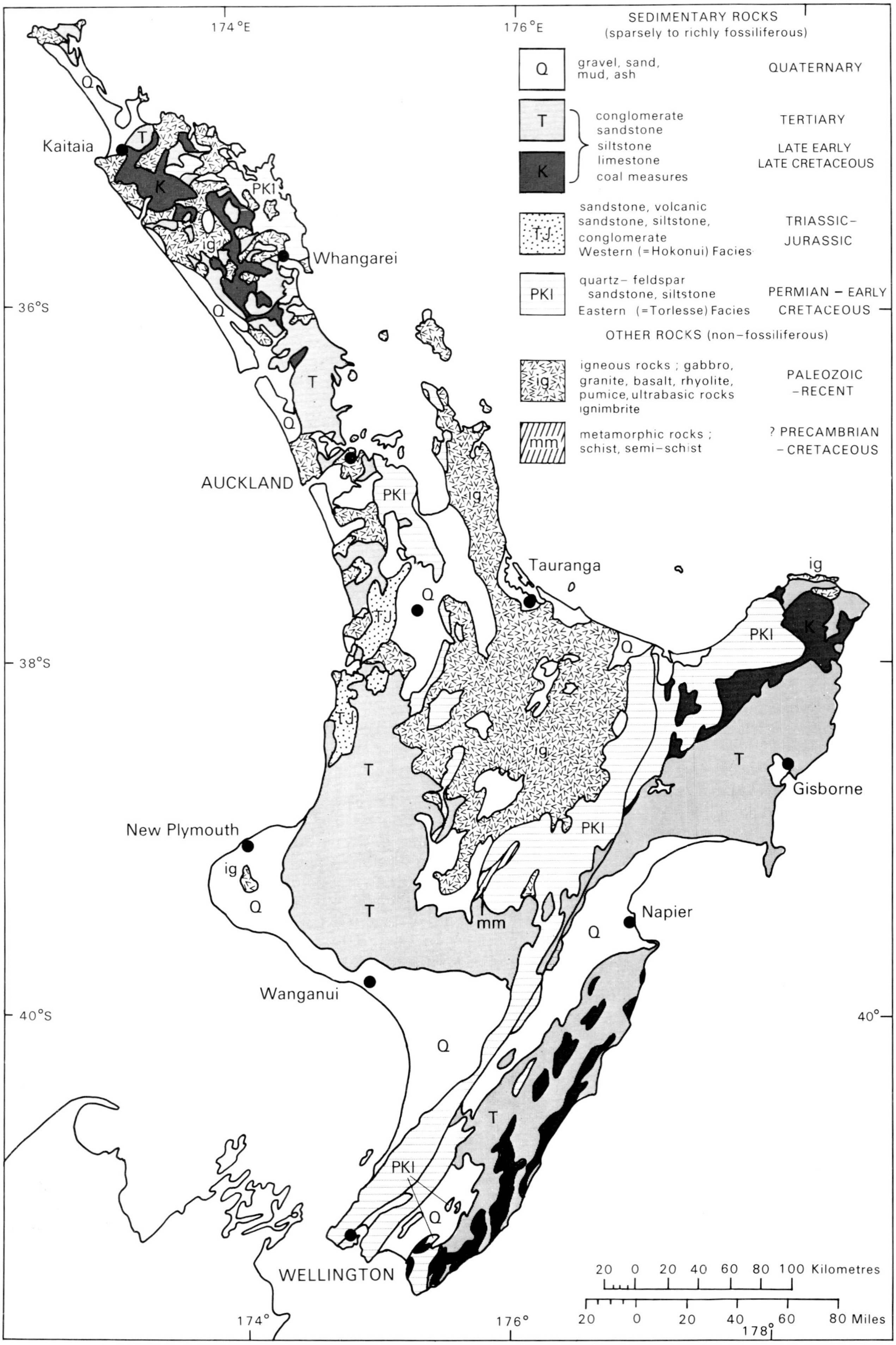

Fig. 1 Simplified geological map of the North Island, New Zealand.

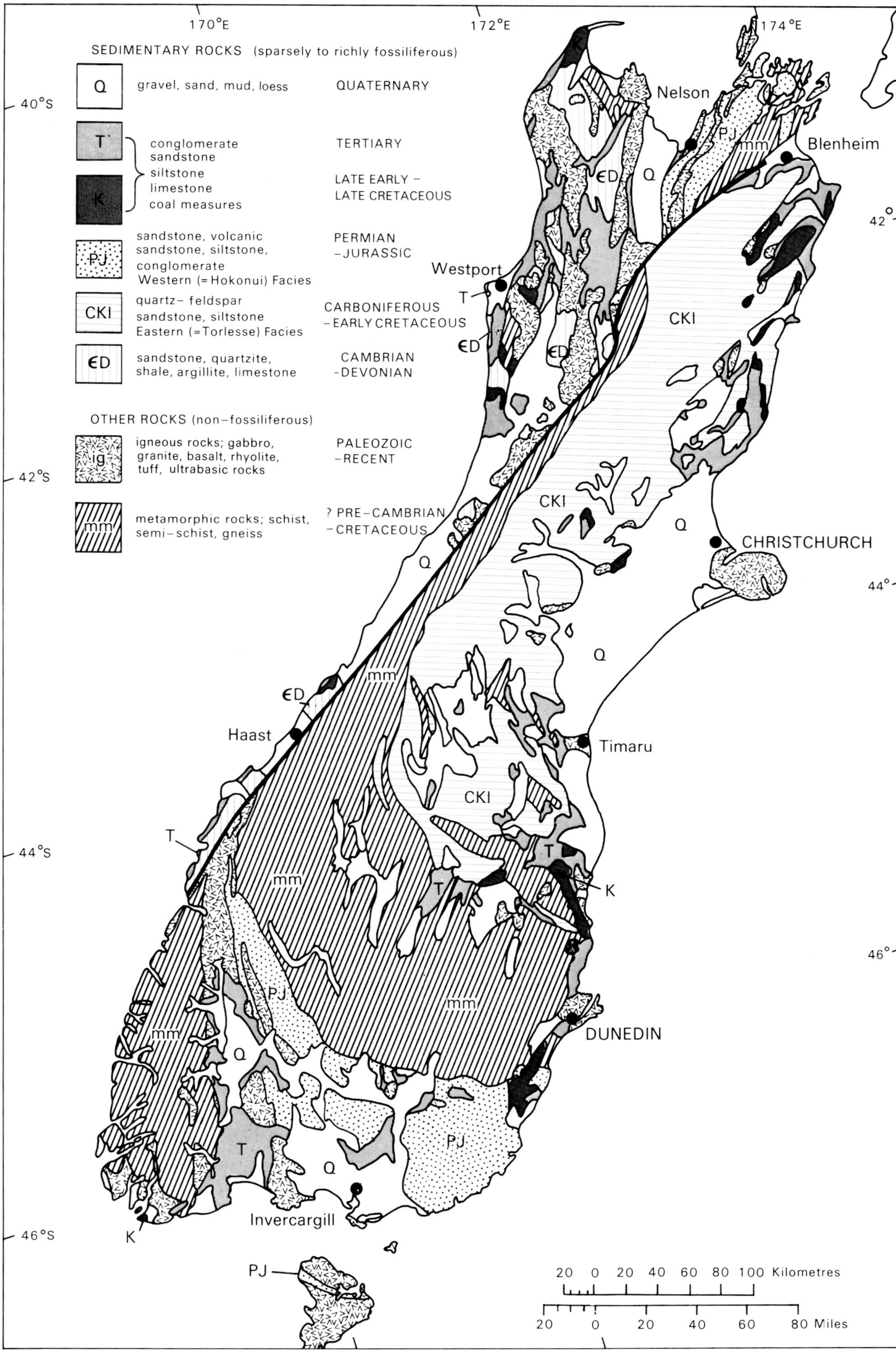

Fig. 2 Simplified geological map of the South Island, New Zealand.

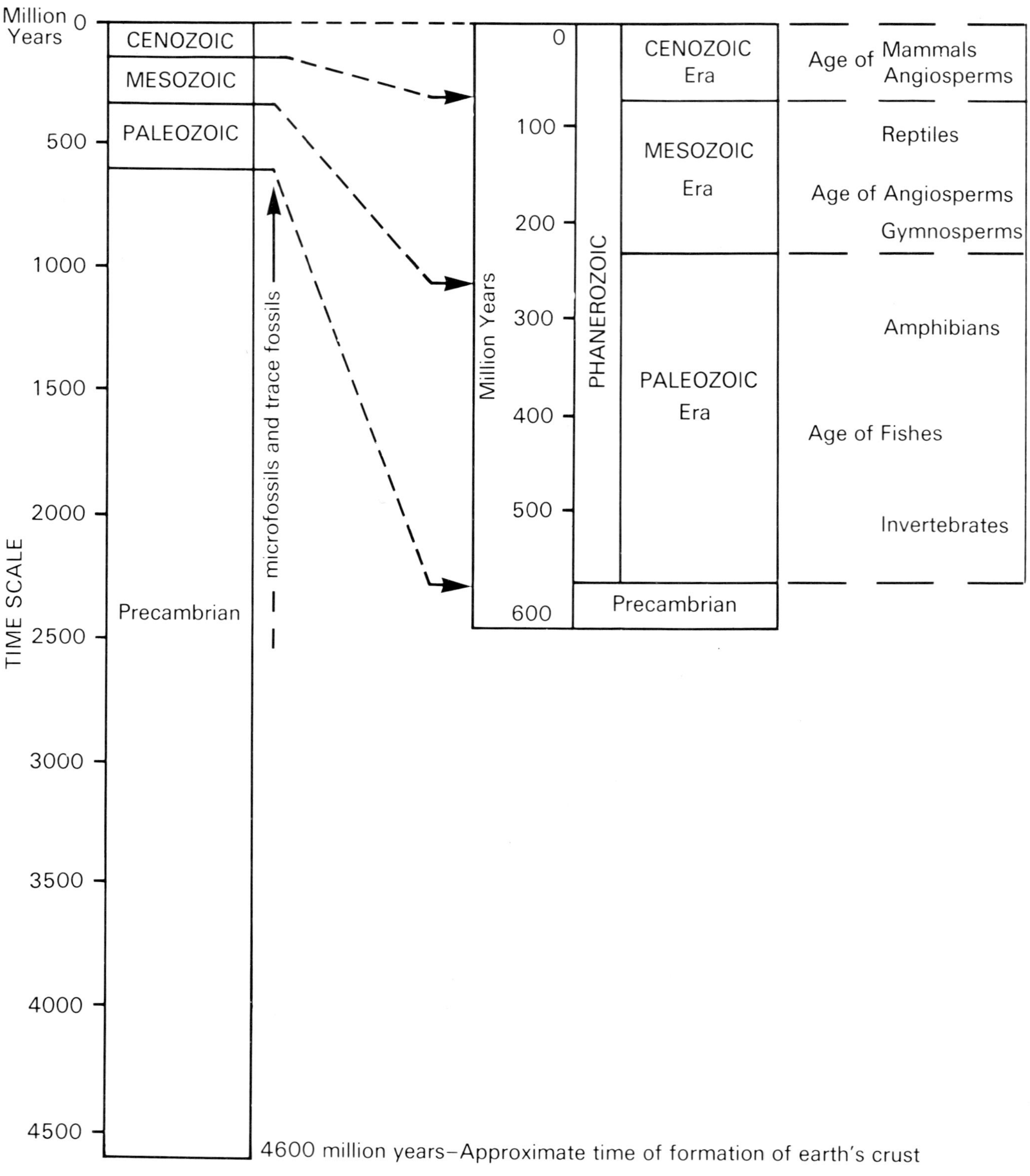

Fig. 3 Geological time scale, indicating the age of major groups of animals and plants. Modified from Suggate *et al.* (1979).

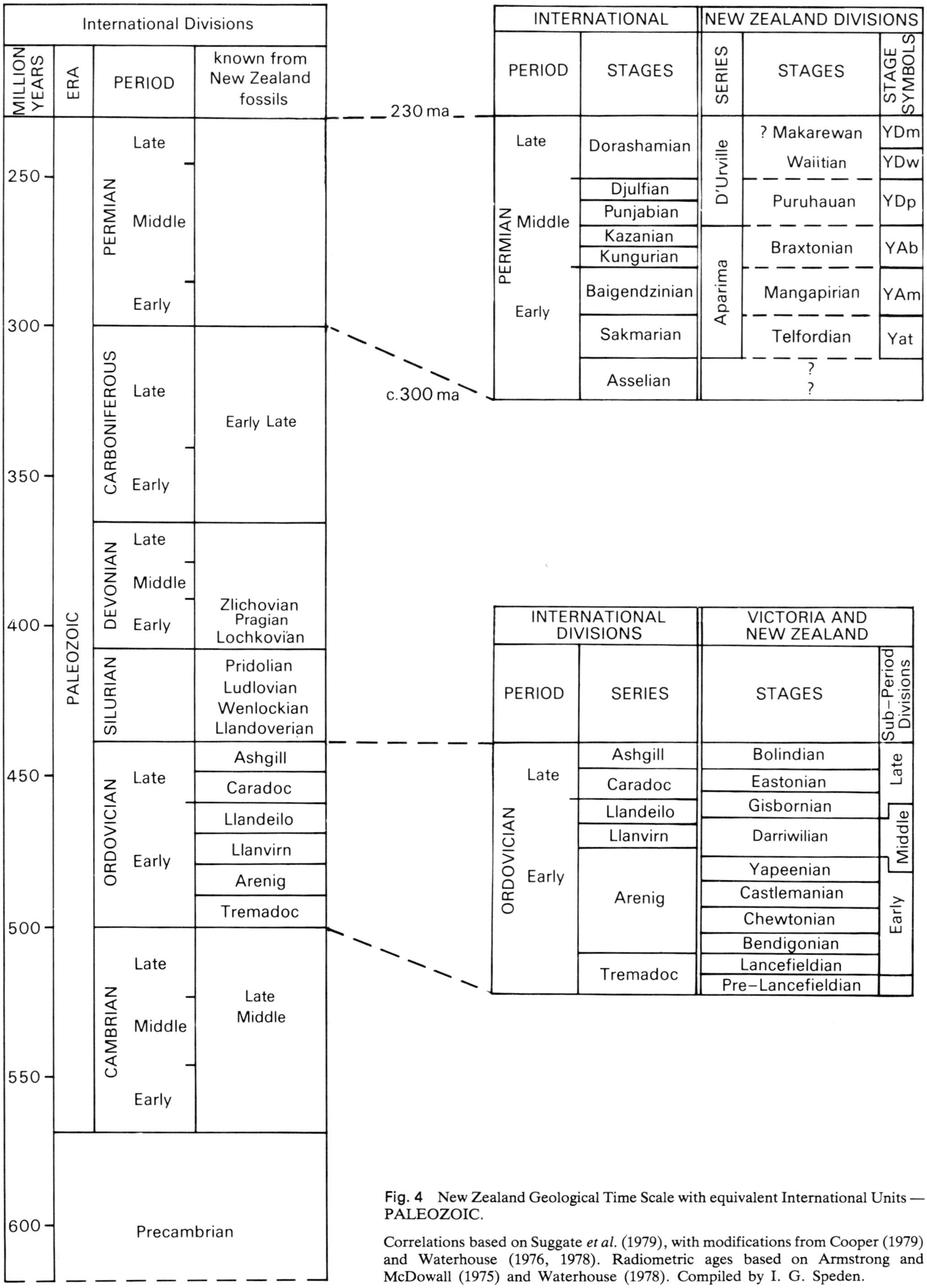

Fig. 4 New Zealand Geological Time Scale with equivalent International Units — PALEOZOIC.

Correlations based on Suggate *et al.* (1979), with modifications from Cooper (1979) and Waterhouse (1976, 1978). Radiometric ages based on Armstrong and McDowall (1975) and Waterhouse (1978). Compiled by I. G. Speden.

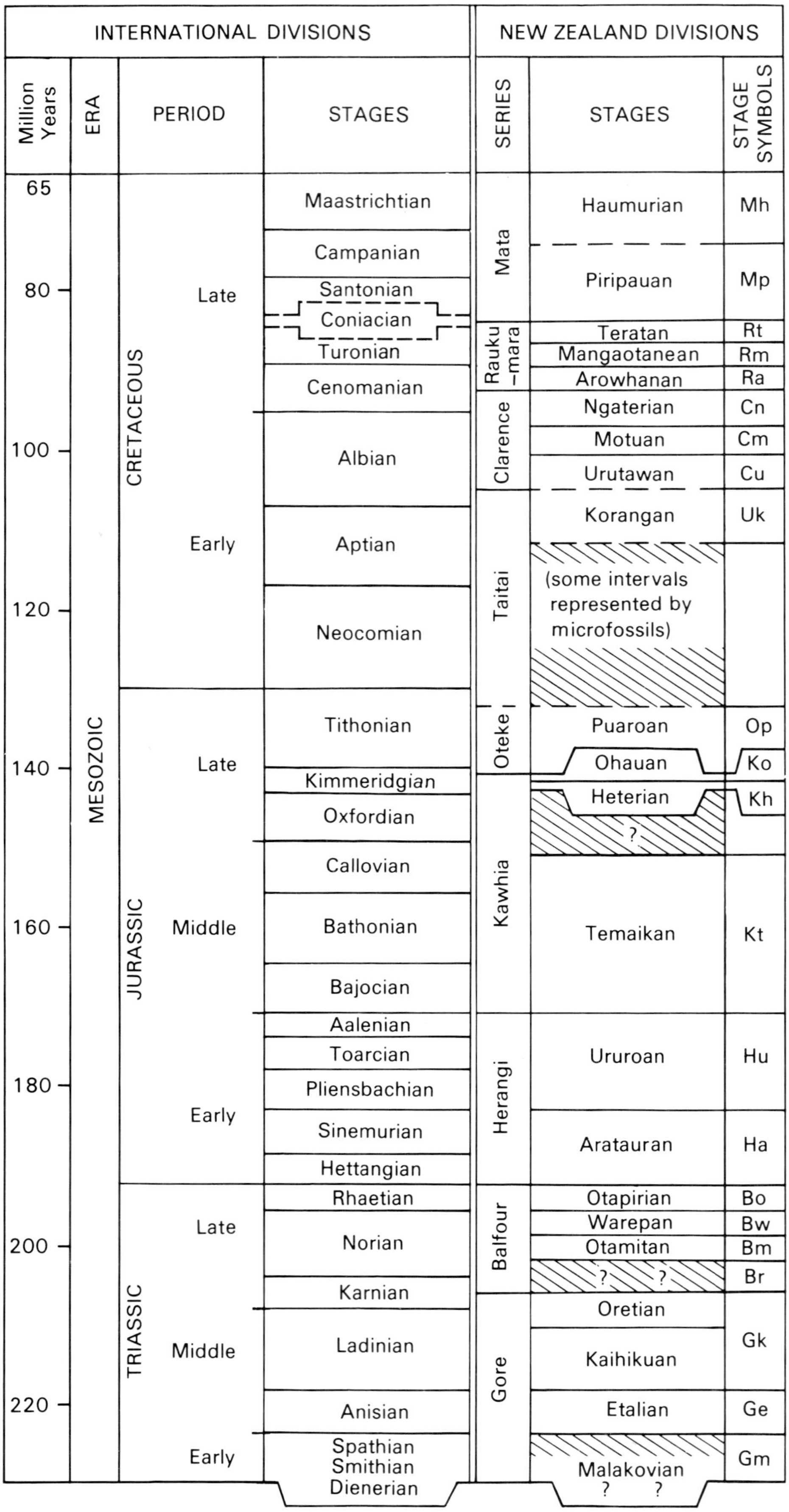

Fig. 5 New Zealand Geological Time Scale with equivalent International Units — MESOZOIC.

Correlations based on Suggate *et al.* (1979), with modifications from Stevens and Speden (1978), and unpublished information by J. D. Campbell. Radiometric ages based on Waterhouse (1978, Triassic), van Hinte (1976, Jurassic), and Odin (1978, Cretaceous). The Dienerian, as a stage or substage, is included by some authors in the Triassic. The placing of the Aalenian Stage in the Lower Jurassic is currently being examined. Compiled by I. G. Speden.

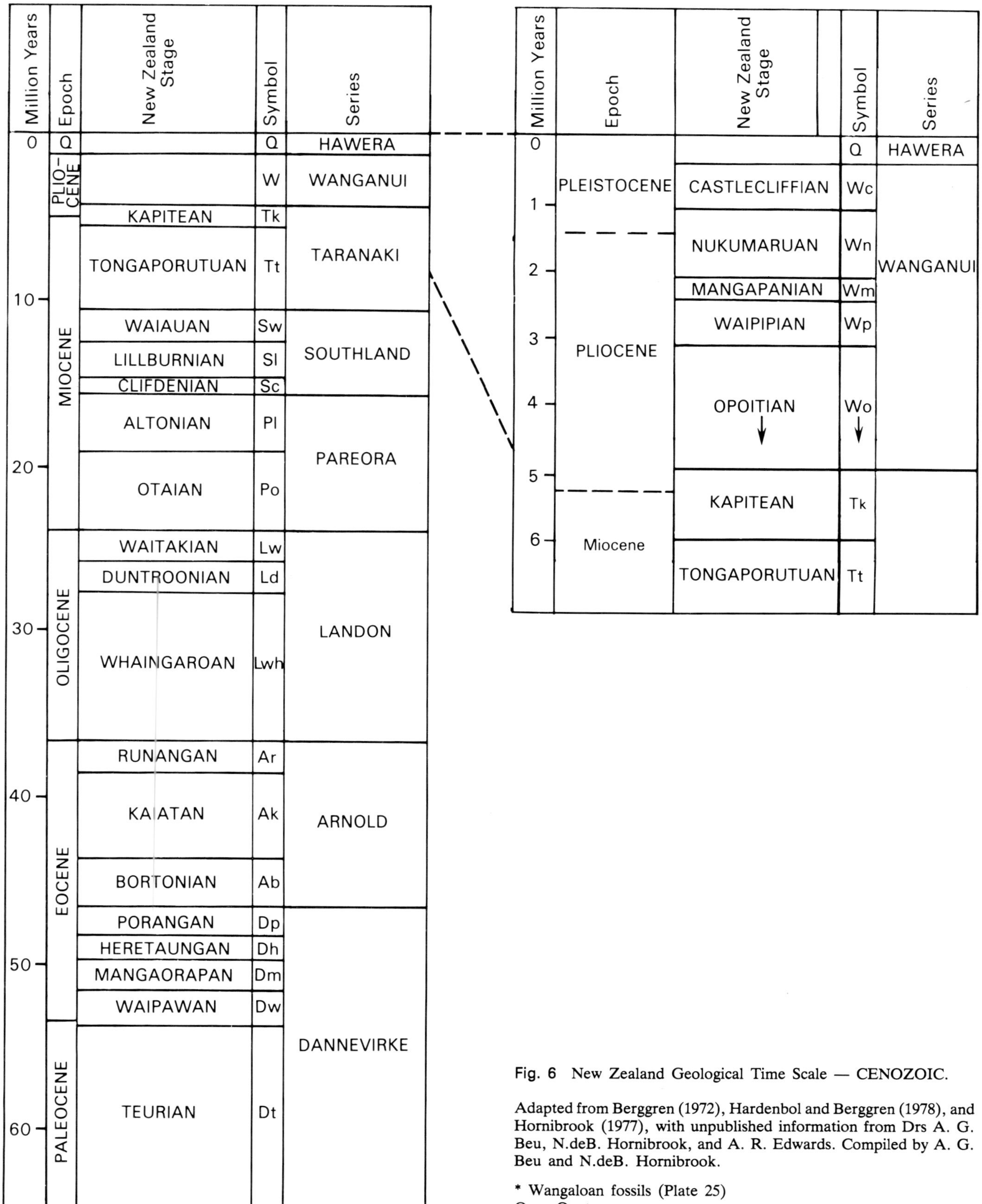

Fig. 6 New Zealand Geological Time Scale — CENOZOIC.

Adapted from Berggren (1972), Hardenbol and Berggren (1978), and Hornibrook (1977), with unpublished information from Drs A. G. Beu, N.deB. Hornibrook, and A. R. Edwards. Compiled by A. G. Beu and N.deB. Hornibrook.

* Wangaloan fossils (Plate 25)
Q = Quaternary

Plate 1

Representative Cambrian and Ordovician trilobites, brachiopods, and conodonts.

a ***Nepea*** sp. Cranidium, late Middle Cambrian. ×2.9.

b ***Dorypyge*** sp. Pygidium, late Middle Cambrian. ×3.5.

c ***Ptychagnostus*** cf. ***aculeatus*** (Angelin). Pygidium, late Middle Cambrian. ×6.3.

d ***Koptura*** sp. Group of 3 pygidia, late Middle Cambrian. ×2.6.

e ***Grandagnostus*** sp. Pygidium, late Middle Cambrian. ×10.

f ***Ptychagnostus*** sp. Cephalon, late Middle Cambrian. ×9.

g ***Micromitra*** sp. Inarticulate brachiopod, late Middle Cambrian. ×22.

h ***Acrothele*** sp. Inarticulate brachiopod, late Middle Cambrian. ×24.

i ***Westergardodina*** sp. Conodont, late Middle Cambrian. ×50.

j ***Hirsutodontus hirsutus*** Miller 1969. Conodont, Tremadoc Series. ×100.

k ***Cordylodus*** cf. ***angulatus*** Pander. Conodont, Tremadoc Series. ×50.

l ***Eoplacognathus*** cf. ***reclinatus*** (Fahraeus). Conodont ambalodiform element. Llanvirn Series. ×35.

m ***Periodon aculeatus*** Hadding. Conodont, periodontiform element. Llanvirn Series. ×50.

n ***Pygodus*** cf. ***anserinus*** Lamont & Lindstrom. Conodont pygodiform element. Upper Llanvirn to Lower Llandeil Series. ×80.

NZGS photos: R. A. Cooper & J. E. Simes.
Compiled: R. A. Cooper & J. E. Simes.

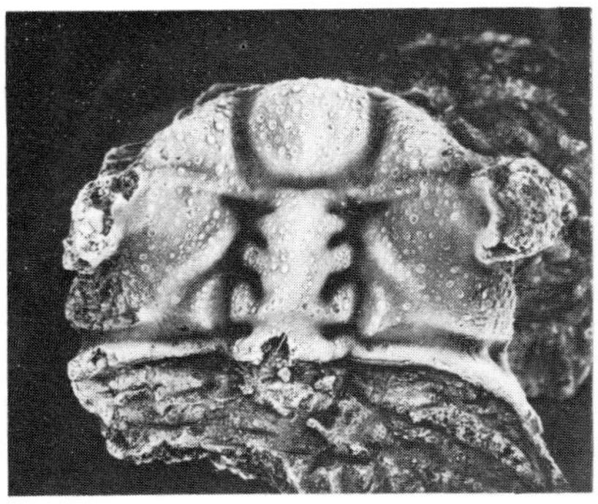

a

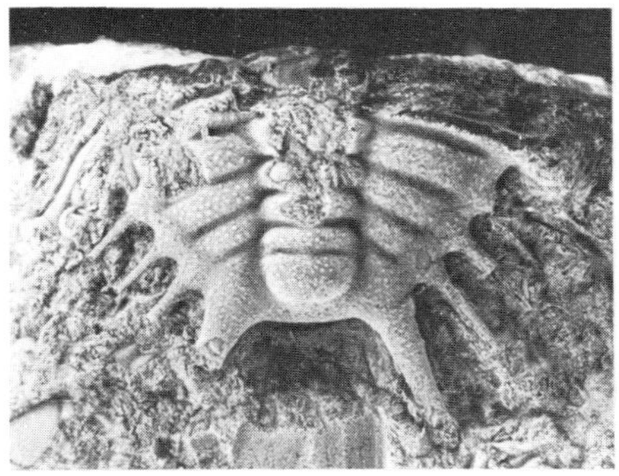

b

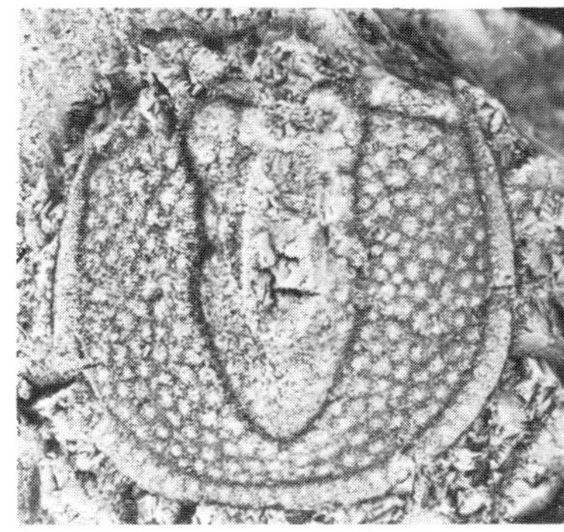

c

d

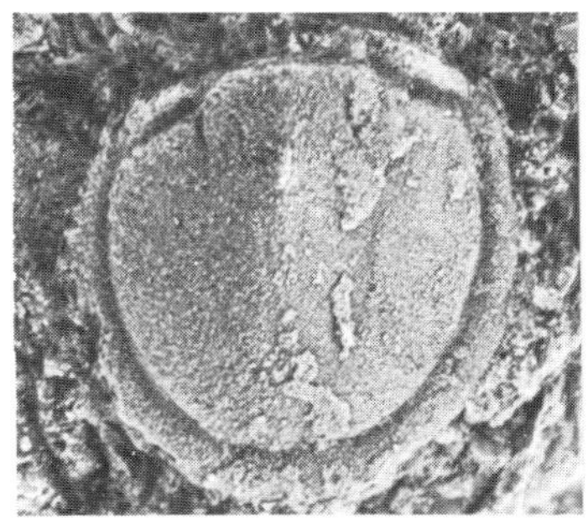

e

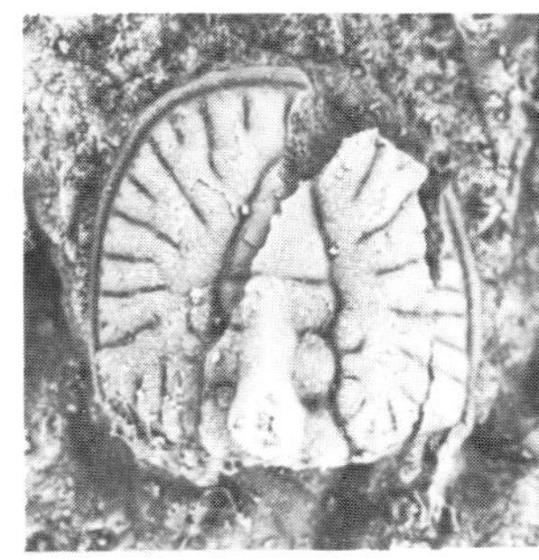

f

g

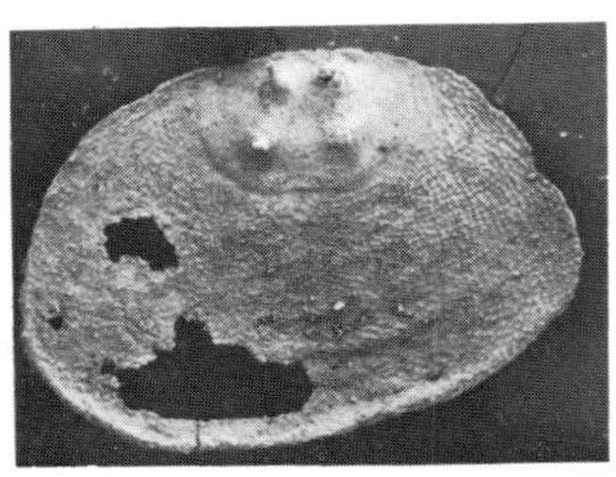

h

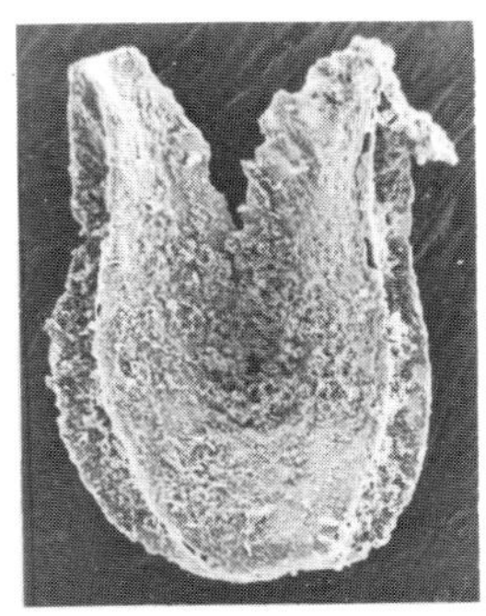

i

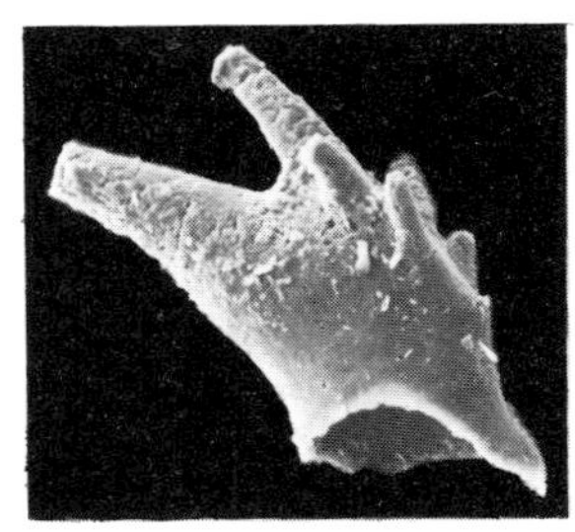

j

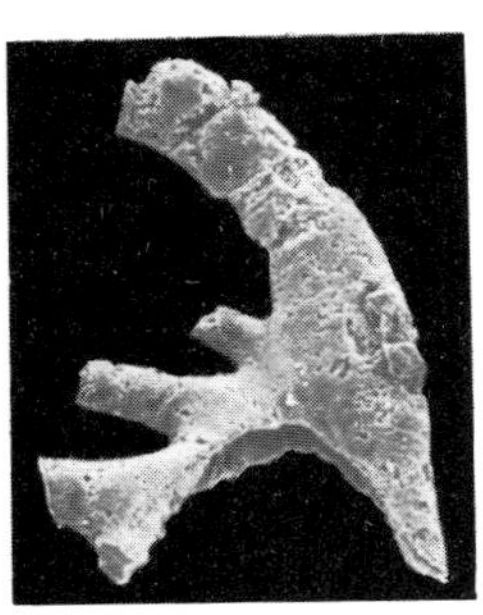

k

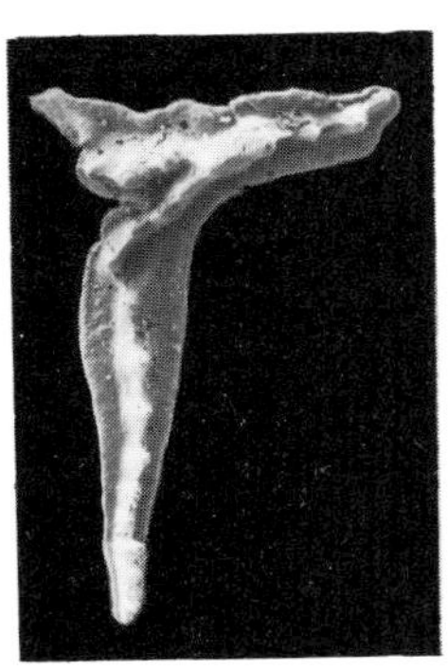

l

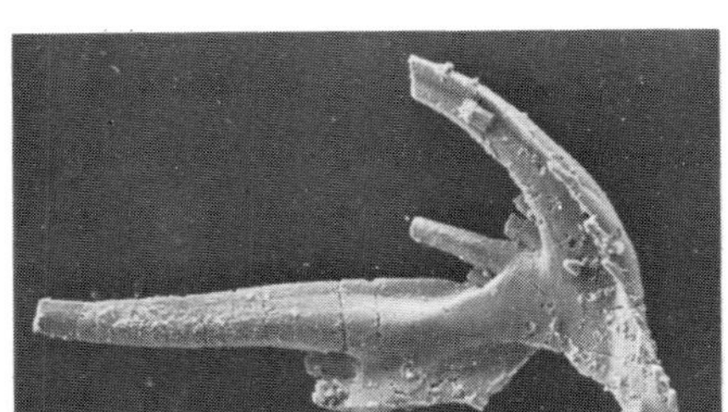

m

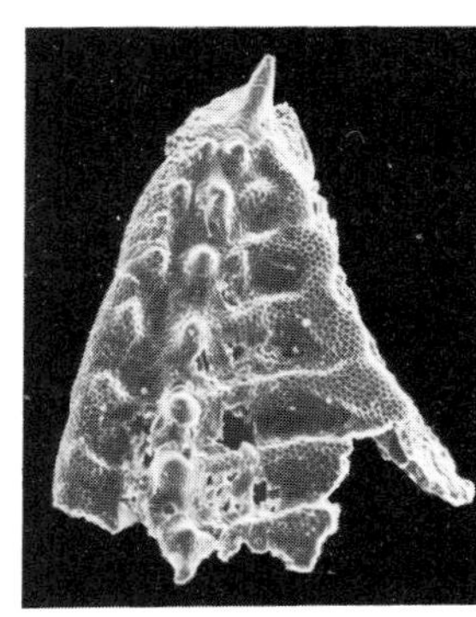

n

Plate 2

Representative Cambrian trilobites and Lower Ordovician graptolites.

a Trilobite limestone with various polymerid trilobites including ***Pianaspis***. Middle Cambrian. ×1.3.

b ***Didymograptus protobifidus*** Elles. Chewtonian (Lower Ordovician). ×1.5.

c ***Tetragraptus fruticosus*** (Hall). Bendigonian (Lower Ordovician). ×1.5.

d ***Isograptus victoriae maximodivergens*** Harris. Yapeenian (Lower Ordovician). ×1.5.

e ***Isograptus caduceus imitatus*** Harris. Castlemainian (Lower Ordovician). ×5.

f ***Isograptus victoriae victoriae*** Harris. Castlemainian (Lower Ordovician). ×2.

g ***Isograptus victoriae lunatus*** Harris. Castlemainian (Lower Ordovician). ×5.

h ***Tetragraptus approximatus*** Nicholson. Lancefieldian (Lower Ordovician). ×2.6.

i ***Adelograptus? antiquus*** (Hall). Lancefieldian (Lower Ordovician). ×5.

NZGS photos: T. R. Ulyatt.
Compiled: R. A. Cooper.

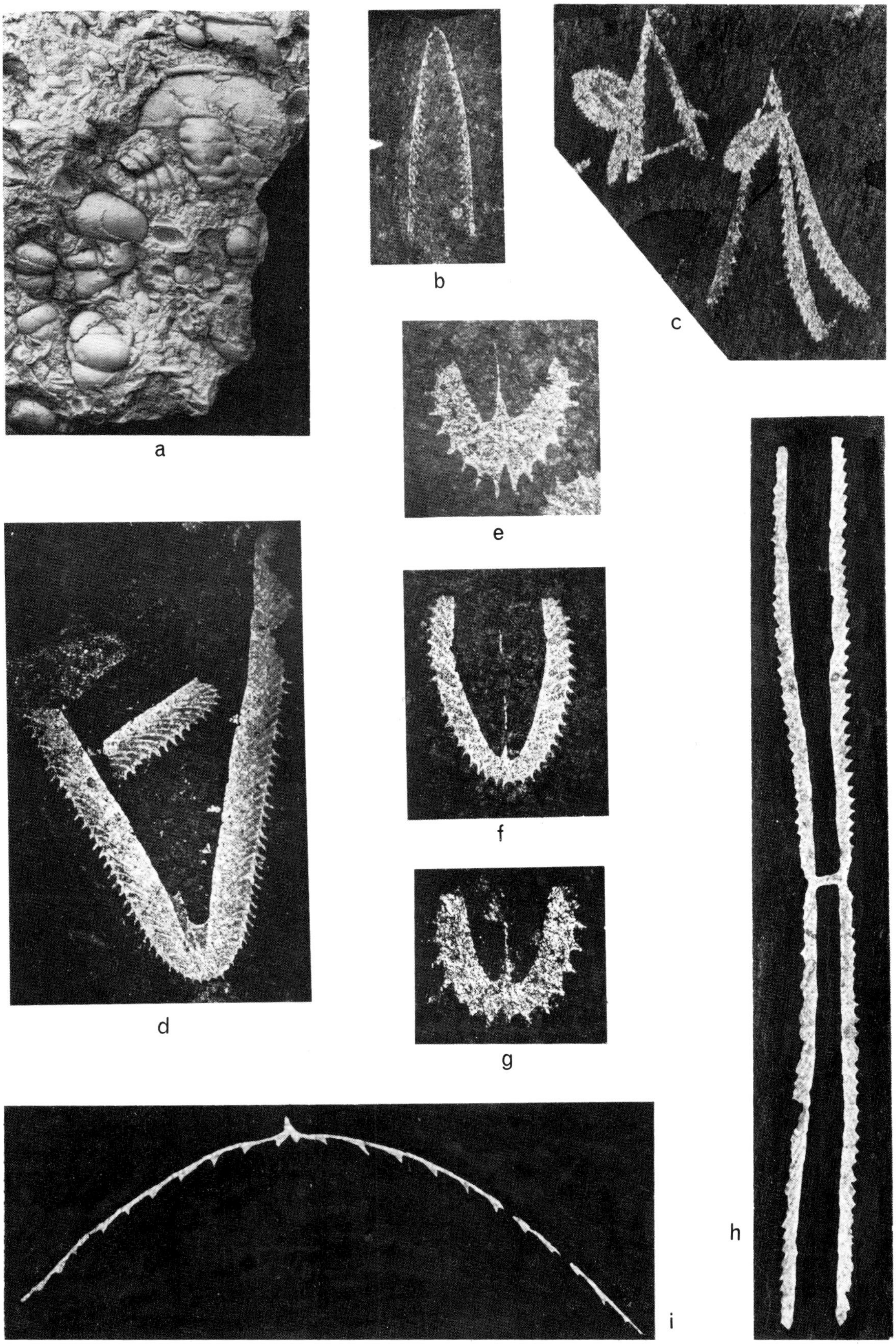
a
b
c
d
e
f
g
h
i

Plate 3

Representative Lower to Upper Ordovician graptolites and trilobite.

a ***Glyptograptus*** cf. ***teretiusculus*** (Hisinger). Upper Darriwilian (Middle Ordovician). ×2.

b ***Cardiograptus morsus*** Harris & Keble. Yapeenian (Lower Ordovician). ×3.5.

c ***Paracardiograptus*** sp. Yapeenian (Lower Ordovician). ×3.

d ***Paraglossograptus tentaculatus*** Harris. Lower Darriwilian (Middle Ordovician). ×3.

e ***Ogygites collingwoodensis*** Reed. Gisbornian (Upper Ordovician). ×1.

f ***Oncograptus upsilon biangulatus*** Harris. Yapeenian (Lower Ordovician). ×3.

g ***Nemagraptus gracilis*** (Hall). Gisbornian (Upper Ordovician). ×1.5.

h ***Orthograptus*** sp. Eastonian (Upper Ordovician). ×2.

NZGS photos: T. R. Ulyatt.
Compiled: R. A. Cooper.

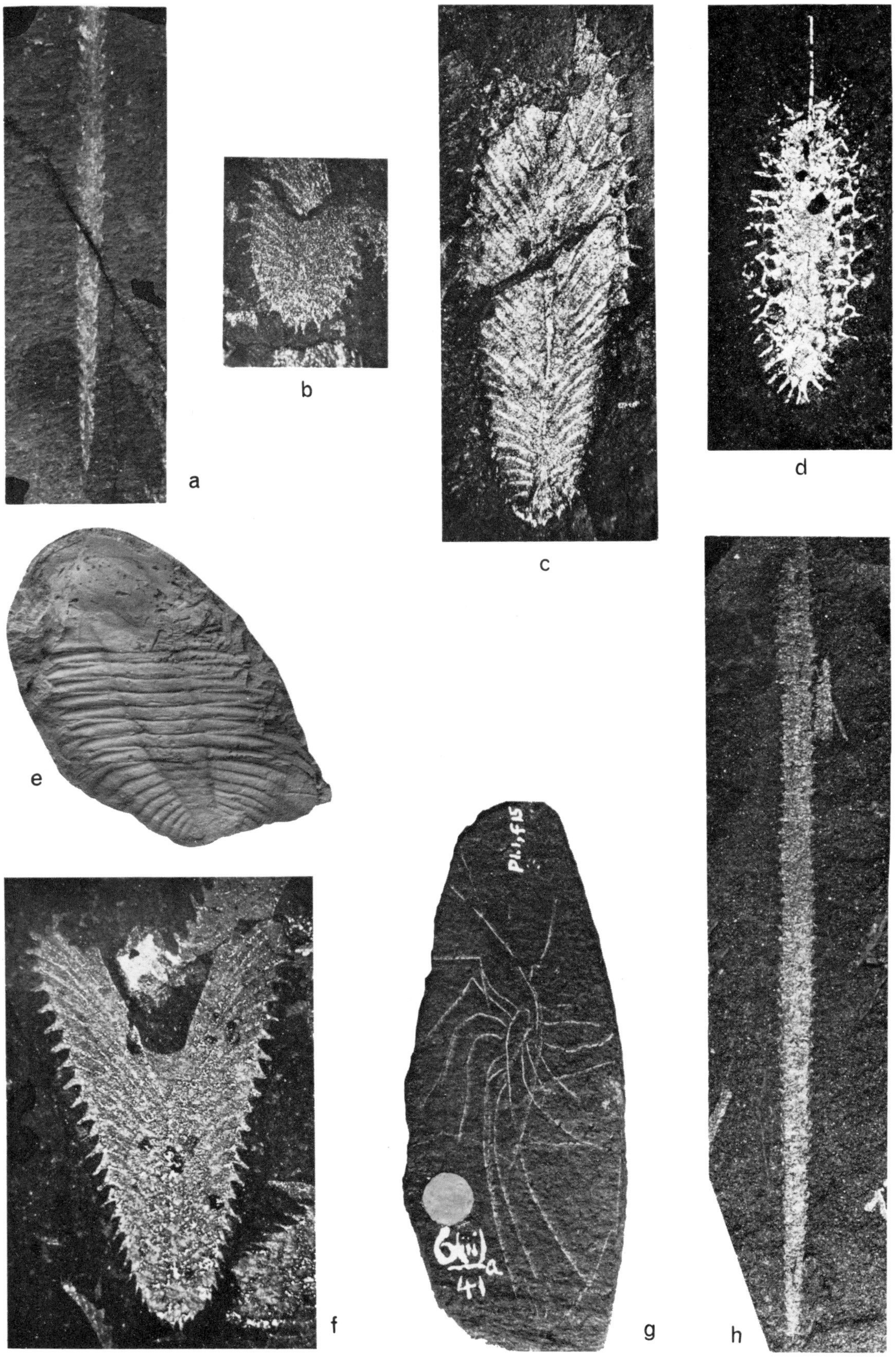
a
b
c
d
e
f
g
h

Plate 4

Representative Lower Devonian fossils.

a ***Ptychopteria (Actinopteria) mackayi*** Fleming. Lower Devonian. ×1.

b ***Fascicostella gervillei*** (Defrance). Lower Devonian. ×2.

c ***Pleurodictyum megastomum*** Dun. Lower Devonian. ×1.5.

d ***Pleurothyrella venusta*** Boucot *et al.* Lower Devonian. ×1.5.

e ***Reeftonella neozelanica*** (Allan). Lower Devonian. ×1.

f ***Hipparionyx minor*** Clarke. Lower Devonian. ×1.

g ***Tipheophyllum bartrumi*** (Allan). Transverse section. Lower Devonian. ×1.8.

h ***Eatonia bithynica*** Paeckelmann & Sieverts. Lower Devonian. ×1.5.

i ***Burmeisteria (Digonus)*** cf. ***expansus*** (Hector). Lower Devonian. ×0.66.

j ***Acrospirifer coxi*** (Allan). Lower Devonian. ×1.

NZGS photos: T. R. Ulyatt.
Compiled: R. A. Cooper.

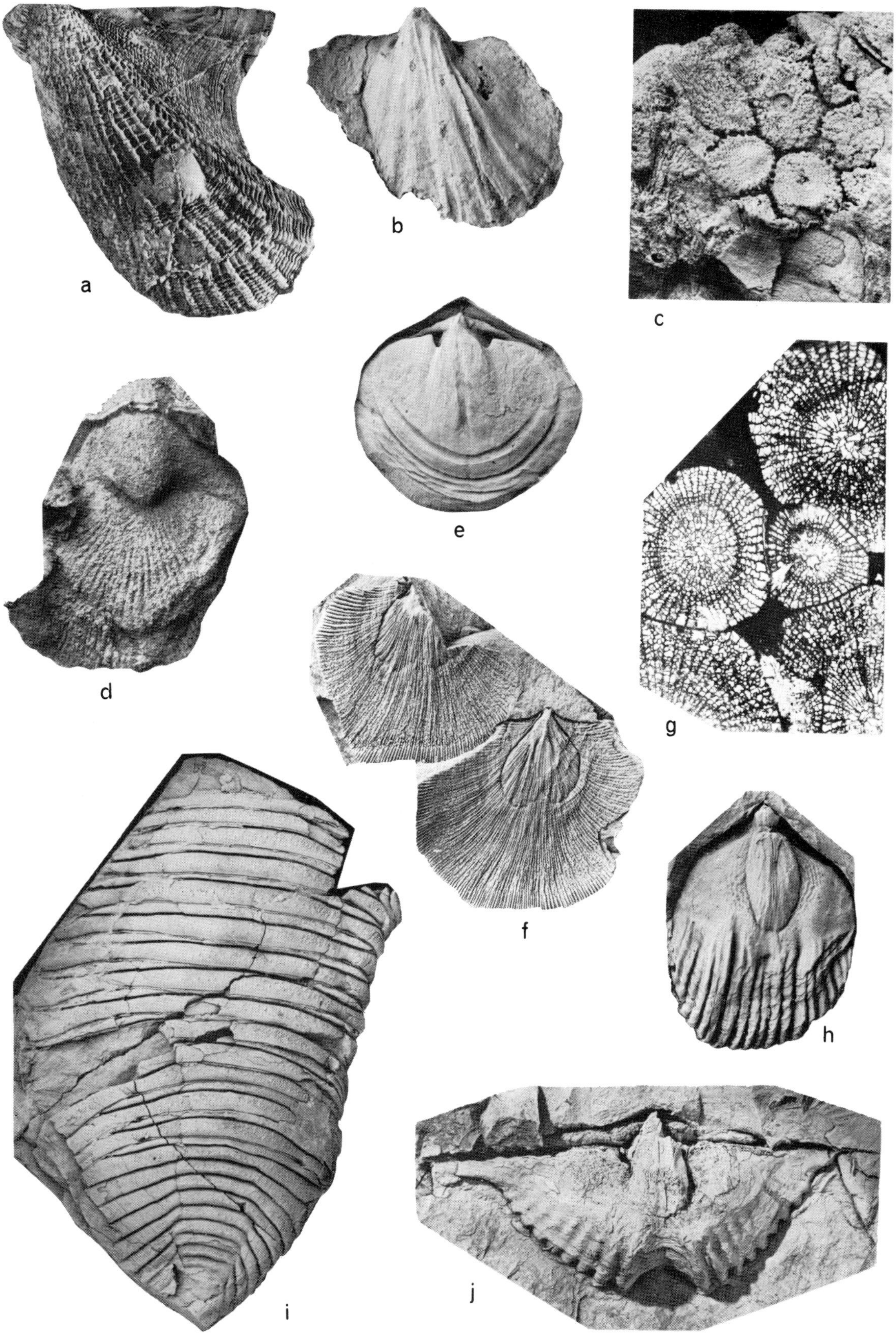
a
b
c
d
e
f
g
h
i
j

Plate 5

Representative Lower and Upper Permian fossils.

1	***Attenuatella altilis*** Waterhouse. Mangapirian Stage. ×4.
2, 3	***Plekonella campbelli*** Waterhouse. Telfordian Stage. 2, ×5; 3, ×2.
4, 5, 11	***Terrakea brachythaerum*** (Morris). Upper Braxtonian Stage. ×2.
6	***Horridonia*** n. sp. Puruhauan Stage. ×1.5.
7, 8	***Psilocamara saginatum*** Waterhouse. Telfordian Stage. ×2.
9, 12	***Mourlonia impressa*** Waterhouse. Pre-Telfordian Stage. ×2.
10, 13	***Aperispirifer wairakiensis*** (Waterhouse). Lower Braxtonian Stage. ×1.
14, 15	***Walnicholsia subcancellata*** (Morris). Upper Braxtonian Stage. ×2.
16, 18	***Martinia adentata*** Waterhouse. Mangapirian Stage. ×1.
17	Bryozoan fragments; ***Fenestella*** sp. to left, ***Protoretepora ampla*** (Lonsdale) to right. Upper Braxtonian Stage. ×4.

NZGS photos: S. N. Beatus.
Compiled: G. R. Stevens.

1
2
3
4
5
6
7
8
9
10
11
12
13
14
15
16
17
18

Plate 6

Representative Upper Permian fossils.

1	***Trabeculatia trabeculum*** (Waterhouse). Puruhauan Stage. ×1.
2, 4, 5	***Tomiopsis mantuanensis*** (Campbell). Upper Braxtonian Stage. ×1.
3	***Atomodesma obliquatum*** Waterhouse. Lower Braxtonian Stage. ×1.
6	***Peruvispira imbricata*** Waterhouse. Upper Braxtonian Stage. ×6.
7	***Myonia compacta*** Waterhouse. Lower and Upper Braxtonian Stages. ×1.
8	***Vacunella oblonga*** Waterhouse. Upper Braxtonian Stage. ×1.
9	***Myonia*** sp. A. Lower Braxtonian Stage. ×1.
10–12	***Aperispirifer lethamensis*** Waterhouse. Upper Braxtonian Stage. 10, ×0.75; 11, ×1; 12, ×0.75.

NZGS photos: S. N. Beatus & D. L. Homer.
Compiled: G. R. Stevens.

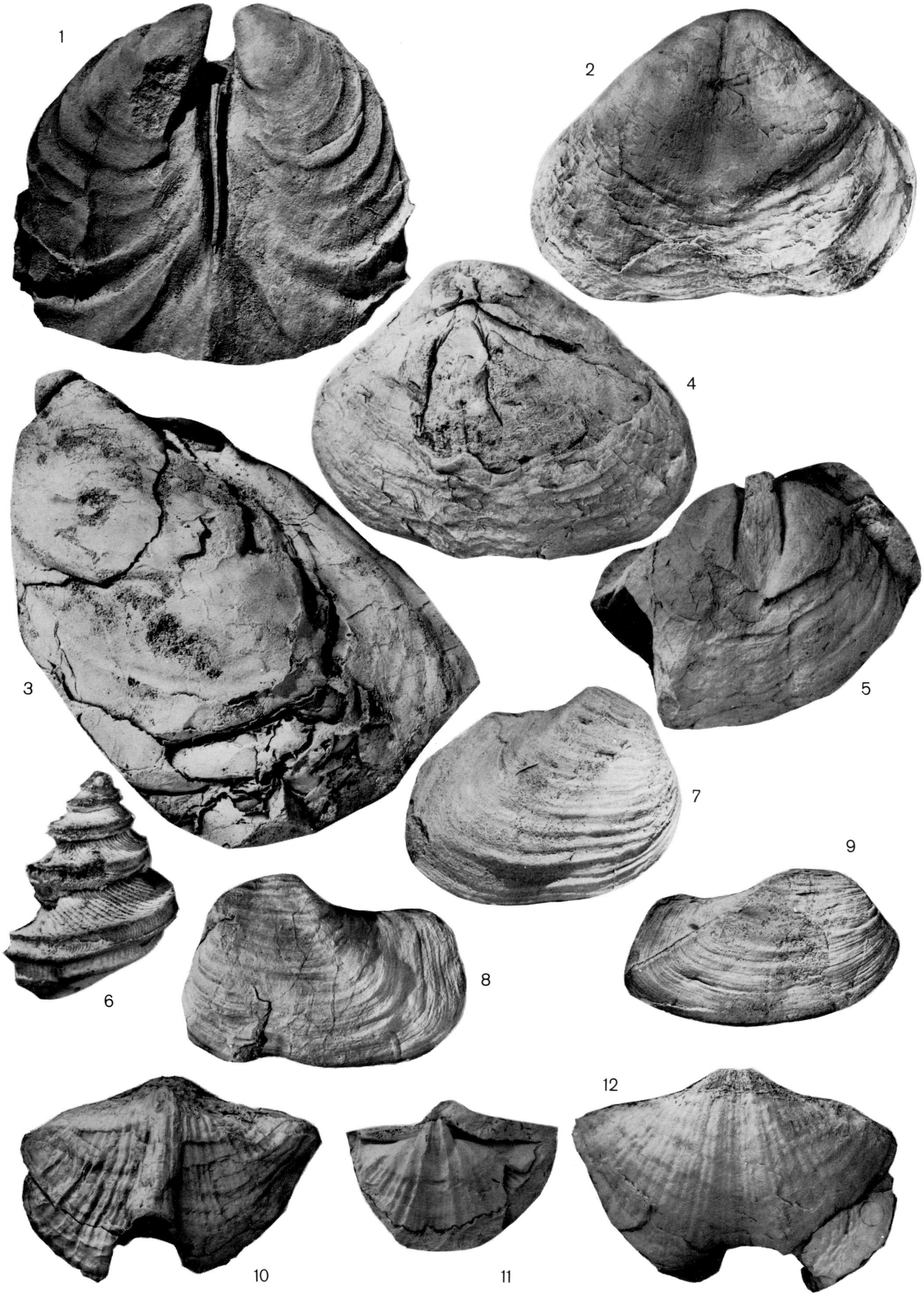
1
2
3
4
5
6
7
8
9
10
11
12

Plate 7

Representative Upper Permian fossils.

1–3	***Plekonella iniquitas*** Waterhouse. Makarewan Stage. ×4.
4, 9	***Durvilleoceras woodmani*** Waterhouse. Puruhauan Stage. ×2.
5, 7	***Etheripecten striatura*** Waterhouse. Puruhauan Stage. ×2.
6, 10	***Capillonia brevisulcus*** (Waterhouse). Puruhauan Stage. ×2.
8	***Attenuatella incurvata*** Waterhouse. Puruhauan Stage. ×4.
11, 13, 14	***Tomiopsis furca*** (Waterhouse). Makarewan Stage. ×1.
12	***Aperispirifer nelsonensis*** (Waterhouse). Waiitian Stage. ×3.
15, 16	***Notospirifer spinosa*** Waterhouse. Upper Braxtonian Stage. ×1.5.
17	***Waagenophyllum novaezelandiae*** Leed. Puruhauan Stage. Transverse section of part of corallum. ×6.
18	***Wentzelella maoria*** Leed. Puruhauan Stage. Transverse section of part of corallum. ×9.

NZGS photos: S. N. Beatus.
R. C. Brazier del.
Compiled: G. R. Stevens.

1
2
3
4
5
7
6
8
9
10
11
12
13
14
15
16
17
18

Plate 8

Endemic New Zealand Permian brachiopods and gastropods.

1–4 ***Coronopsis vagrans*** Waterhouse. Upper Braxtonian Stage. ×2.

5–7 ***Wairakiella rostrata*** Waterhouse. Makarewan Stage. ×4.

8–10 ***Spinomartinia spinosa*** Waterhouse. Puruhauan Stage. ×3.

NZGS photos: S. N. Beatus.
R. C. Brazier del.
Compiled: G. R. Stevens.

1
2
3
4
5
6
7
8
9
10

Plate 9

Representative Triassic fossils.

1, 2 ***Stenopopanoceras coombsi*** (Kummel). Malakovian Stage. ×1.5.

3 ***Retzia reticulata*** Wilckens. Otamitan Stage. ×2.

4 ***Retzia morganiana*** Wilckens. Otamitan Stage. ×2.

5 ***Amphipopanoceras fraseri*** (Browne). Etalian Stage. ×1.

6 ***Owenites*** cf. ***koeneni*** Hyatt & Smith. Malakovian Stage. ×1.

7 ***Amphipopanoceras tepungai*** (Browne). Etalian Stage. ×1.

8 ***Flemingites*** cf. ***lidacensis*** Welter. Malakovian Stage. ×

9 ***Patella nelsonensis*** Trechmann. Kaihikuan Stage. ×1.

10, 12, 13 ***Mentzeliopsis spinosa*** Trechmann. Kaihikuan Stage. ×1.

11, 15, 16, 17 ***Daonella apteryx*** Marwick. Kaihikuan Stage. Natural internal moulds. ×1.

14 ***Leiophyllites marshalli*** Browne. Etalian Stage. ×1.

18, 19 ***Spiriferina trechmanni*** (Wilckens). Kaihikuan Stage. ×1.

20, 21 ***"Rhynchonella" zealandica*** (Trechmann). Kaihikuan Stage. Natural internal moulds. ×2.

22 ***Dielasma zealandica*** Trechmann. Kaihikuan Stage. Natural internal mould. ×1.

23, 24 ***Oretia coxi*** Marwick. Oretian Stage. Natural internal moulds. ×1.

NZGS photos: D. L. Homer.
R. C. Brazier del.
Compiled: G. R. Stevens.

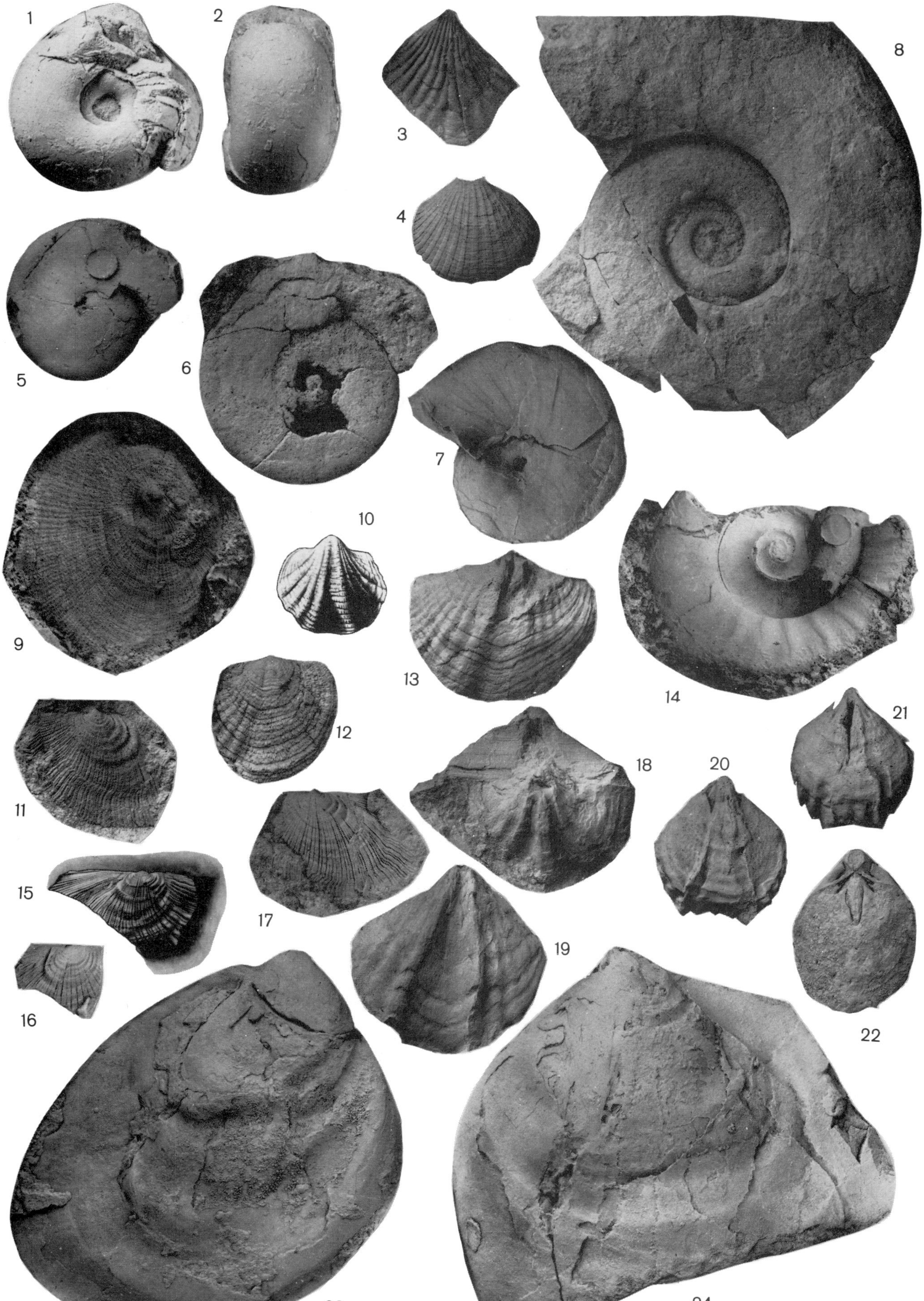
1
2
3
4
5
6
7
8
9
10
11
12
13
14
15
16
17
18
19
20
21
22
23
24

Plate 10

Representative Triassic fossils.

1, 2 ***Proclydonautilus mandevillei*** (Marshall). Otamitan Stage. 2 shows ornament on portion of outer whorl. 1, ×0.5; 2, ×1.

3, 4 ***Halobia hochstetteri*** Mojsisovics. Oretian and Otamitan Stages. ×1.

5 ***Poroa arata*** (Trechmann). Oretian and Otamitan Stages. ×1.

6, 8 ***Maoritrigonia nuggetensis*** (Trechmann). Oretian Stage. 6, ×1; 8, ×0.9.

7, 11 ***Hokonuia limaeformis*** (Trechmann). Otamitan Stage. 7, ×1.5; 11, ×1.

9, 13 ***Mysidioptera riceae*** Waterhouse. Otamitan Stage. ×2.

10 ***Raha spectabilis*** (Trechmann). Otamitan Stage. ×1.

12 ***Cucullaea wellmani*** Marwick. Oretian Stage. ×1.5.

14 ***Psioidiella conjuncta*** (Hector). Oretian Stage. ×1.

NZGS photos: D. L. Homer.
R. C. Brazier del.
Compiled: G. R. Stevens.

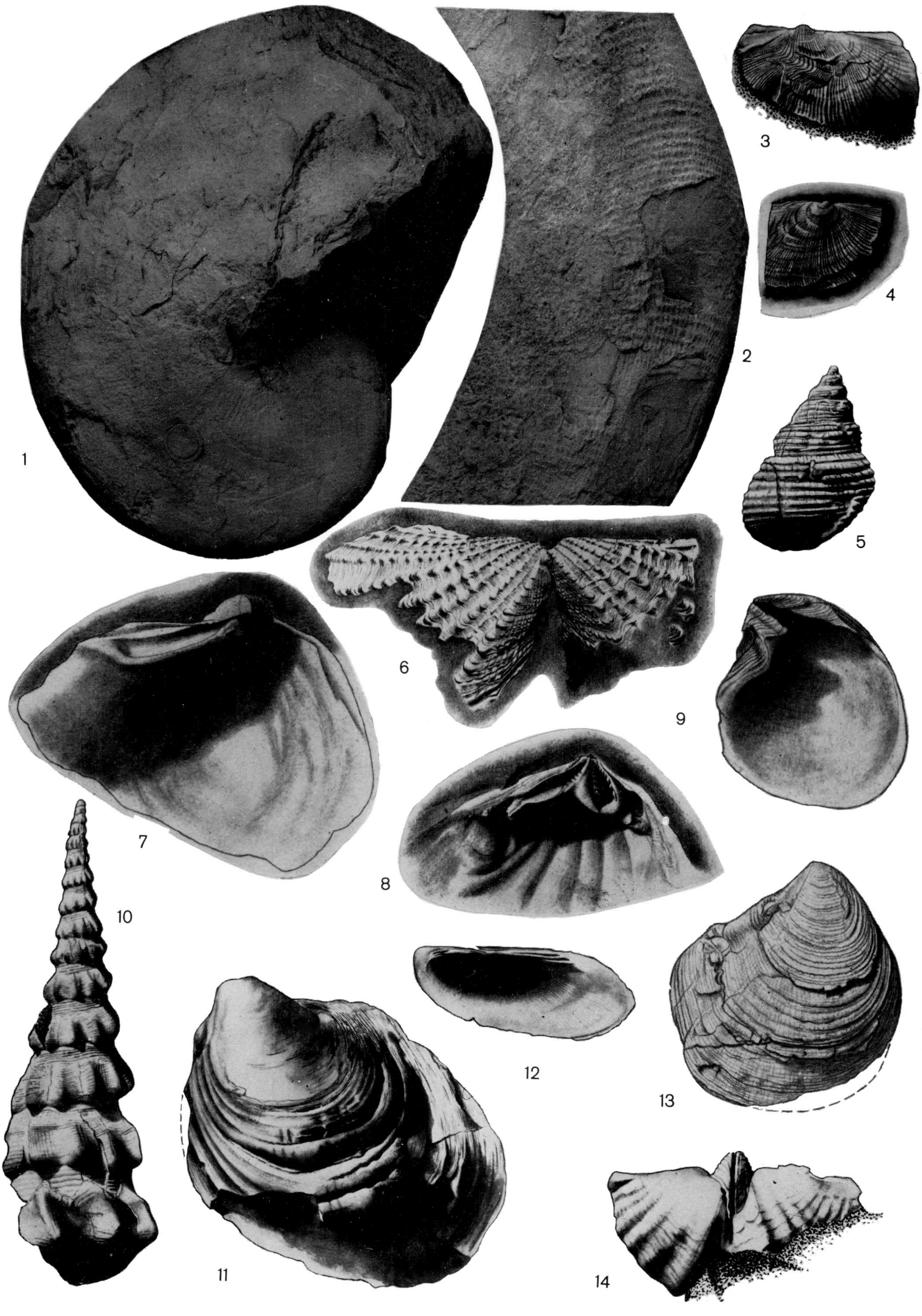
1
2
3
4
5
6
7
8
9
10
11
12
13
14

Plate 11

Representative Triassic fossils.

1, 5 ***Clavigera bisulcata*** Hector. Otapirian Stage. ×1.

2, 6 ***Clavigera tumida*** Hector. Otapirian Stage. Natural internal mould. ×1.

3, 4, ,7 ***Otapiria dissimilis*** (Cox). Otapirian Stage. 3, ×1; 4, ×1.5; 7, ×1.4.

8 ***Monotis (Entomonotis) richmondiana richmondiana*** Zittel. Warepan Stage. Natural internal moulds. ×1.

9 ***Rastelligera elongata*** Hector. Otapirian Stage. Natural internal mould. ×1.

10, 13 ***Monotis (Entomonotis) richmondiana aucticostata*** (Trechmann). Warepan Stage. ×1.

11 ***Rastelligera gypaeetus*** (Trechmann). Otapirian Stage. External mould. ×1.

12, 14 ***Mentzelia kawhiana*** Trechmann. Otapirian Stage. Natural internal moulds. ×1.

15 ***Rhacophyllites debilis*** (Hauer). Otamitan Stage. ×0.5.

16 ***Kalentera marwicki*** Grant-Mackie. Otapirian Stage. ×1.5.

17, 19 ***Pleurotomaria (Sisenna) hectori*** Trechmann. Otamitan Stage. ×1.

18 ***Monotis (Maorimonotis) calvata*** Marwick. Warepan Stage. ×1.

20 ***Maoritrigonia*** aff. ***pirouteti*** (Avias). Otamitan Stage ×0.75.

21 ***Maoritrigonia heslingtonensis*** (Trechmann). Otamitan Stage. ×6.

22 ***"Minetrigonia"*** n.sp. Otapirian Stage. ×1.

NZGS photos: D. L. Homer.
R. C. Brazier del.
Compiled: G. R. Stevens.

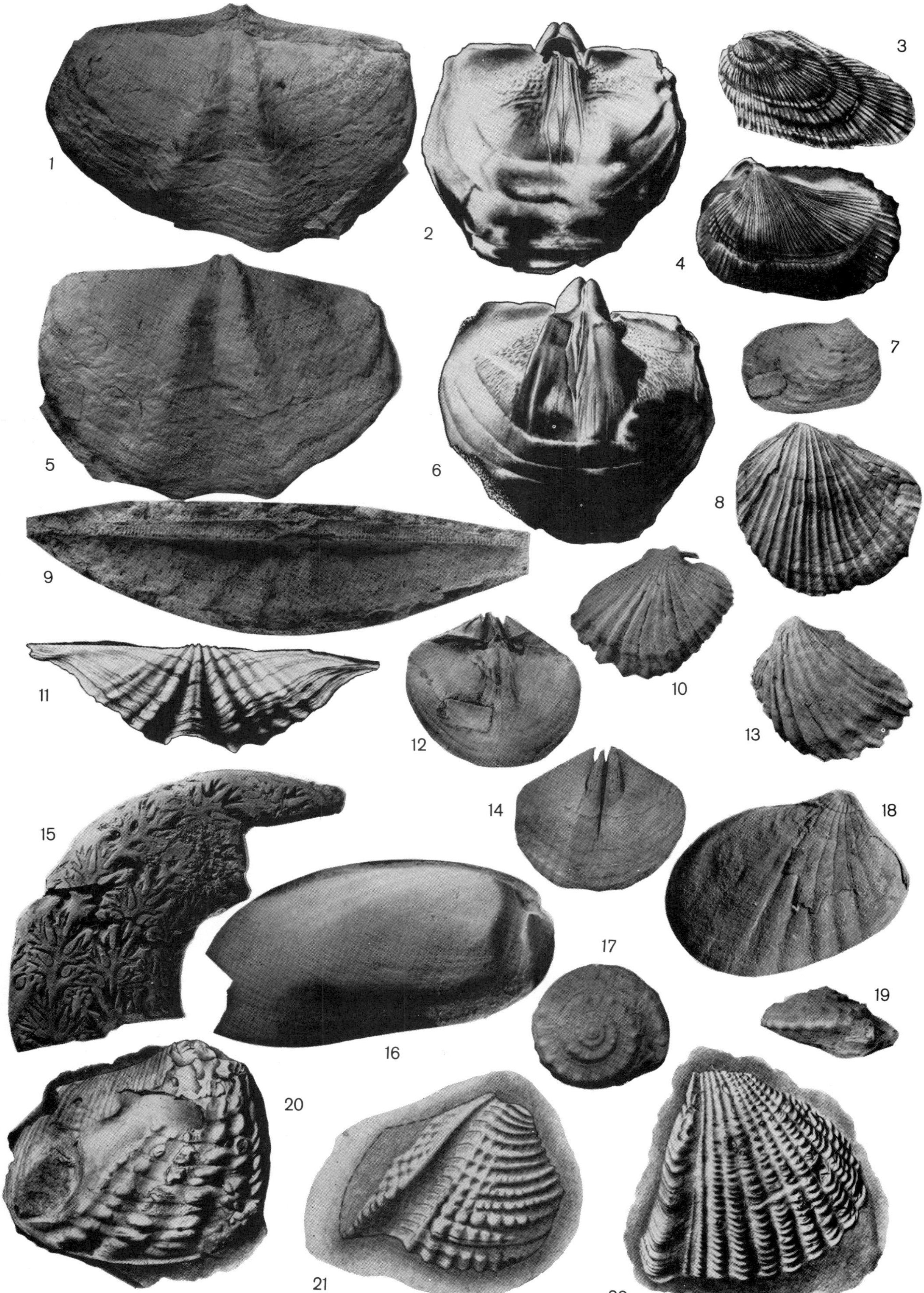
1
2
3
4
5
6
7
8
9
10
11
12
13
14
15
16
17
18
19
20
21
22

Plate 12

Representative Southwest Pacific Triassic brachiopods and molluscs.

1, 2, 5, 6	***Agonisca corbiensis*** Fleming. Kaihikuan Stage. ×1.
3, 7, 8, 11	***Alipunctifera kaihikuana*** (Trechmann). Kaihikuan Stage. 3, ×1; 7, ×0.8; 8, ×1; 11, ×1.
4, 13	***Manticula problematica*** (Zittel). Otamitan Stage. ×1.
9, 10, 15	***Mellarium nodulosum*** Waterhouse. Etalian Stage. ×1.5.
12, 16	***Worthenia phillipsi*** Waterhouse. Etalian Stage. ×3.5.
14, 17	***Kamupena greggi*** Speden. Kaihikuan Stage. ×1.
18, 19	***Torastarte bensoni*** Marwick. Otapirian Stage. ×1.
20–22	***Athyris kaihikuana*** Trechmann. Kaihikuan Stage. ×1.
23, 25–27	***Praegonia coombsi*** Fleming. Kaihikuan Stage. ×1.
24	***Balantioselena gairi*** Speden. Kaihikuan Stage. ×3.
28, 29	***Triaphorus zealandicus*** (Trechmann). Otamitan Stage. ×1.

NZGS photos: S. N. Beatus & D. L. Homer.
R. C. Brazier del.
Compiled: G. R. Stevens.

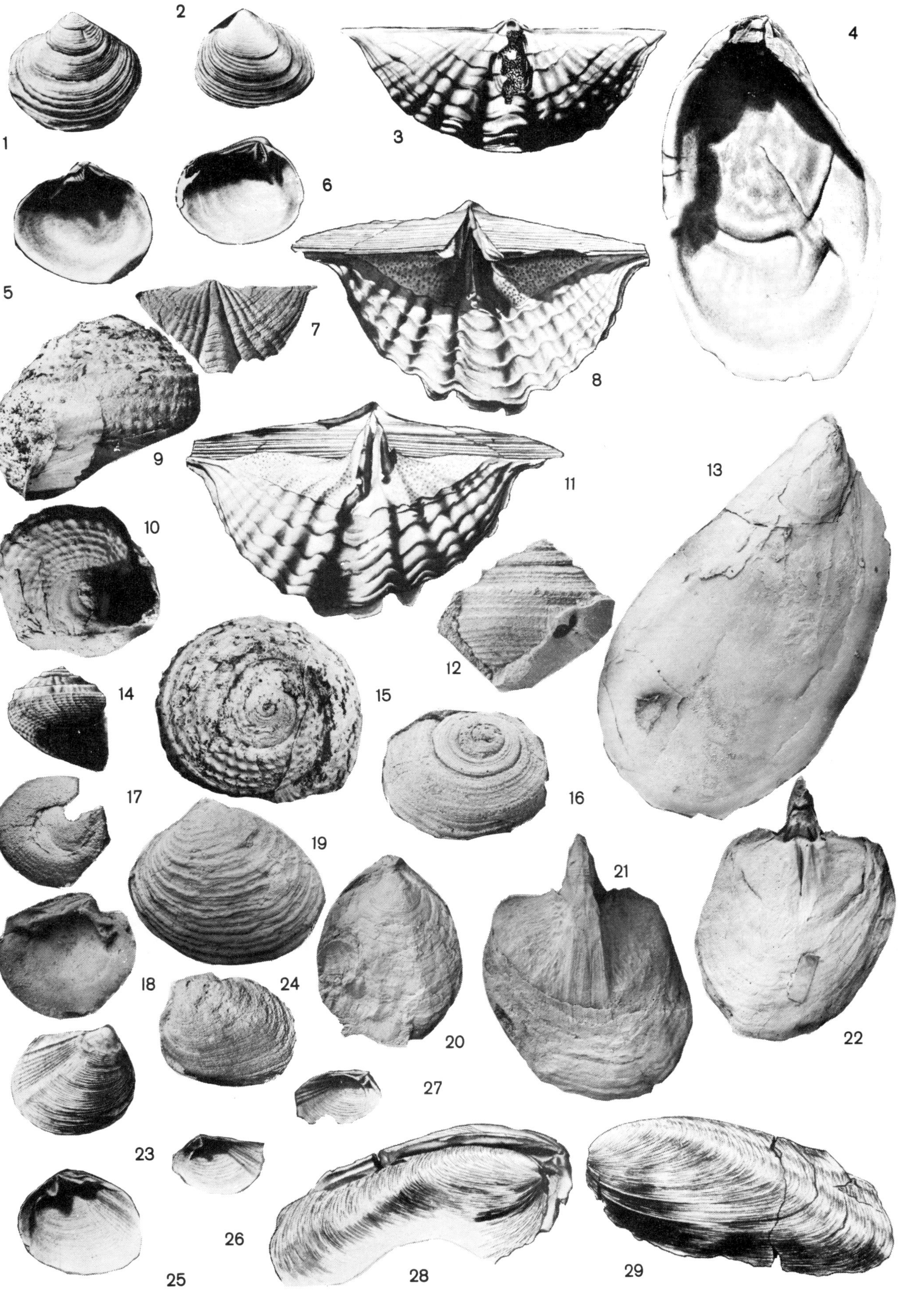
1
2
3
4
5
6
7
8
9
10
11
12
13
14
15
16
17
18
19
20
21
22
23
24
25
26
27
28
29

Plate 13

Representative Lower Jurassic fossils.

1	***Waehneroceras*** n.sp. Aratauran Stage. ×1.
2, 7	***Chlamys (Camptochlamys) wunschae*** Marwick. Ururoan Stage. ×1.
3	***Pseudolimea inanis*** Marwick. Ururoan Stage. ×2.
4, 8	***Otapiria marshalli*** (Trechmann). Aratauran Stage. ×1.
5	***Dactylioceras*** cf. ***anginum*** (Reinecke). Upper Ururoan Stage. ×1.
6	***Harpoceras*** cf. ***falcifer*** (Sowerby). Ururoan Stage. ×1.
9	***Entolium fossatum*** Marwick. Aratauran Stage. ×1.
10	***Pleuromya urnula*** Marwick. Upper Ururoan Stage. ×2.
11	***Tancredia allani*** Marwick. Upper Temaikan Stage. ×1.5.
12, 13	***Pleurotomaria (Perotrochus)*** n.sp. Lower Temaikan Stage. ×1.
14, 15	***Brachybelus zieteni*** (Werner). Lower Temaikan Stage. ×1.
16	***Belemnopsis mackayi*** Stevens Lower Temaikan Stage. ×1.
17, 22	***Parainoceramus martini*** Speden. Upper Ururoan Stage. 17, ×3; 22, ×2.
18	***Pseudolimea fida*** Marwick. Aratauran Stage. ×2.
19	***Trigonia*** (s. str.) n.sp. Temaikan Stage. ×2.
20, 21	***"Rhynchonella" bartrumi*** Marwick. Upper Ururoan Stage. ×2.
23, 24	***Inoceramus ururoaensis*** Speden. Upper Ururoan Stage. 23, ×1; 24, ×2.
25	***Variamussium clamosseum*** Marwick. Lower Temaikan Stage. ×2.
26	***Myophorella*** n.sp. Temaikan Stage. ×1.
27	***Meleagrinella*** cf. ***echinata*** (Smith). Temaikan Stage. ×0.75.
28	***Vaugonia*** n.sp. Temaikan Stage. ×2.
29	***Orthotrigonia*** n.sp. Temaikan Stage. ×1.

NZGS photos: D. L. Homer.
R. C. Brazier. del.
Compiled: G. R. Stevens.

1
2
3
4
5
6
7
8
9
10
11
12
13
14
15
16
17
18
19
20
21
22
23
24
25
26
27
28
29

Plate 14

Representative Middle and Upper Jurassic fossils.

1 ***Camptonectes*** cf. ***laminatus*** (Sowerby). Temaikan Stage. ×1.

2 ***Macrocephalites (Kamptokephalites)*** cf. ***beta-gamma*** (Boehm). Upper Temaikan Stage. ×1.

3, 4 ***Inoceramus marwicki*** Speden. Upper Temaikan Stage. 3, ×1; 4, ×1.5.

5, 6 ***Conodicoelites flemingi*** Stevens. Heterian Stage. ×1.

7, 8 ***Haastina haastiana*** (Wilckens). Upper Temaikan Stage. 7, ×0.5; 8, ×1.

9 ***Vaugonia kawhiana*** (Trechmann). Heterian Stage. ×2.

10, 11 ***Pleuromya milleformis*** Marwick. Temaikan Stage. ×1.

12 ***Kossmatia*** cf. ***desmidoptycha*** Uhlig. Heterian Stage. ×0.75.

13, 15, 21 ***Malayomaorica malayomaorica*** (Krumbeck). Heterian and Ohauan Stages. 13, ×1; 15, ×1; 21, ×1.5.

14 ***Paraboliceras*** aff. ***sabineanum*** (Oppel). Ohauan Stage. ×0.75.

16, 17 ***Kutchithyris hendersoni*** Marwick. Heterian Stage. ×1.

18 ***Pinna kawhiana*** Marwick. Heterian Stage. ×1.

19, 20 ***Cerithinella*** sp. Heterian Stage. 19, ×1.5; 20, ×1.

22 ***Inoceramus brownei*** Marwick. Lower Temaikan Stage. ×1.

23 ***Grammatodon (Indogrammatodon) taylori*** Marwick. Heterian Stage. ×1.

24, 26 ***Cryptorhynchia kawhiana*** (Trechmann). Heterian Stage. ×2.

25 ***Astarte morgani*** (Trechmann). Heterian Stage. ×1.

27 ***Hibolithes catlinensis*** (Hector). Upper Temaikan Stage. ×1.

28 ***Myophorella*** n.sp. Ohauan and Heterian Stages. ×1.

29 ***Belemnopsis aucklandica trechmanni*** Stevens. Ohauan Stage. ×1.

30 ***Belemnopsis keari*** Stevens. Heterian Stage. ×1.

NZGS photos: D. L. Homer.
R. C. Brazier del.
Compiled: G. R. Stevens.

1
2
3
4
5
6
7
8
9
10
11
12
13
14
15
16
17
18
19
20
21
22
23
24
25
26
27
28
29
30

Plate 15

Representative Middle and Upper Jurassic fossils.

1 ***Taeniopteris crassinervis*** (Feistmantel). Upper Temaikan Stage. ×0.5.

2–5 ***Buchia*** aff. ***subpallasi*** (Krumbeck). ?Puaroan Stage. ×1.

6, 7 ***Buchia*** aff. ***misolica*** (Krumbeck). Puaroan Stage. ×1.5.

8–12 ***Buchia plicata*** (Zittel). Puaroan Stage. ×1.

13–15, 19, 20 ***Buchia hochstetteri*** Fleming. Puaroan Stage. ×1.

16 ***Hibolithes arkelli*** Stevens. Puaroan Stage. ×1.

17 ***Hibolithes marwicki marwicki*** Stevens. Puaroan Stage. ×1.

18 ***Aulacosphinctoides brownei*** (Marshall). Puaroan Stage. ×0.75.

NZGS photos: D. L. Homer.
R. C. Brazier del.
Compiled: G. R. Stevens.

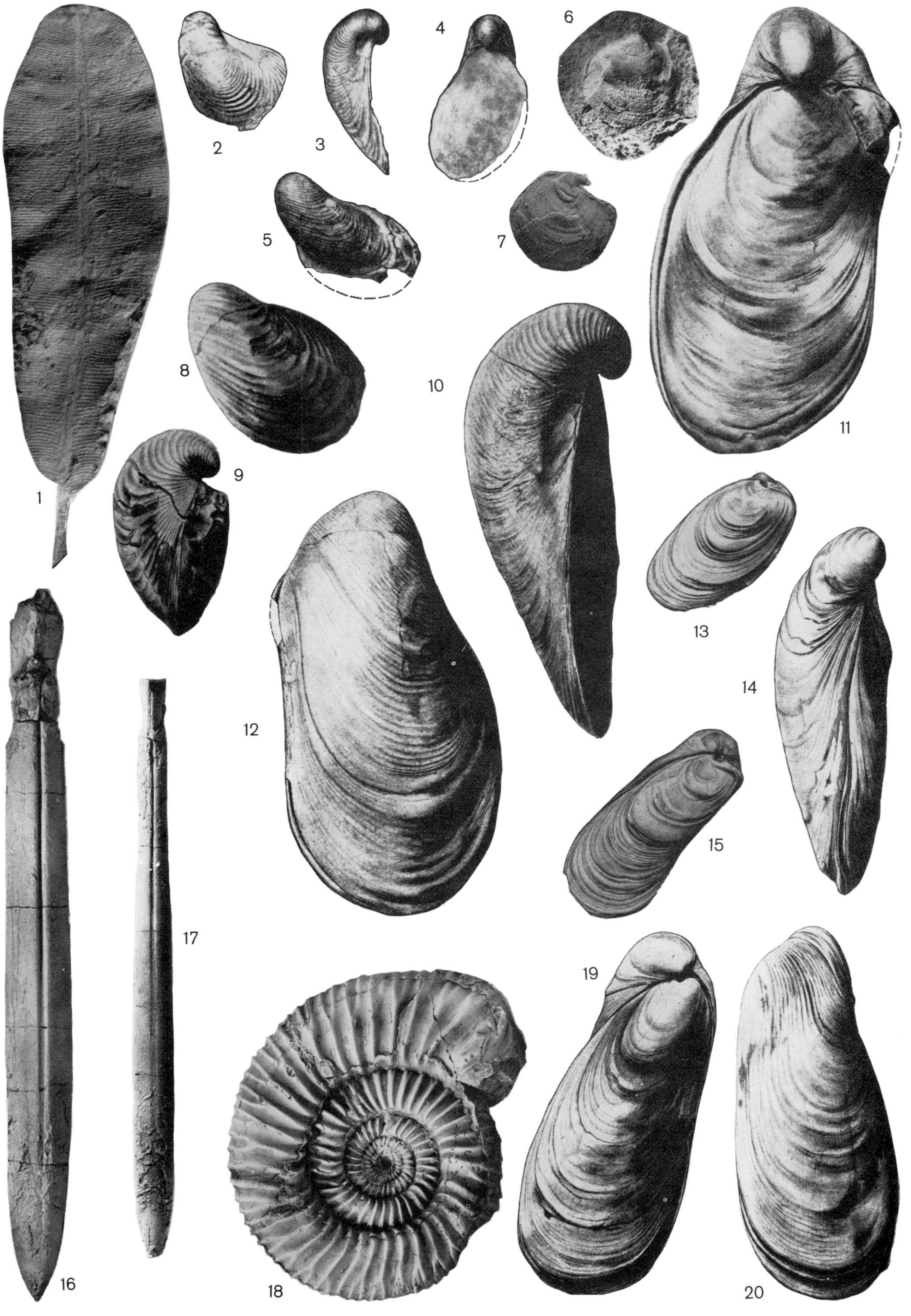
1
2
3
4
5
6
7
8
9
10
11
12
13
14
15
16
17
18
19
20

Plate 16

Representative Upper Jurassic fossils.

1 ***Sphenopteris*** sp. Puaroan Stage. ×2.

2 ***Anopaea*** n.sp. Puaroan Stage. ×1.

3 ***Microphyllopteris pectinata*** Arber ex Hector. Puaroan Stage. ×2.

4 ***Grammatodon (Indogrammatodon) fyfei*** Marwick. Puaroan Stage. ×1.

5 ***Uhligites motutaranus*** (Boehm). Puaroan Stage. ×0.75.

6 ***Psilotrigonia*** n.sp. Puaroan Stage. ×1.

7 ***Myophorella*** n.sp. Puaroan Stage. ×1.

8, 9, 11 ***Otapiria masoni*** Marwick. Puaroan Stage. ×1.

10 ***Nucula (Palaeonucula) putiensis*** Marwick. Puaroan Stage. ×2.

12 ***Torquatisphinctes*** n.sp. Puaroan Stage. ×0.5.

13 ***Inoceramus*** aff. ***everesti*** Oppel. Puaroan Stage. ×1.

14, 15 ***Belemnopsis aucklandica aucklandica*** (Hochstetter). Puaroan Stage. ×1.

NZGS photos: D. L. Homer.
R. C. Brazier del.
Compiled: G. R. Stevens.

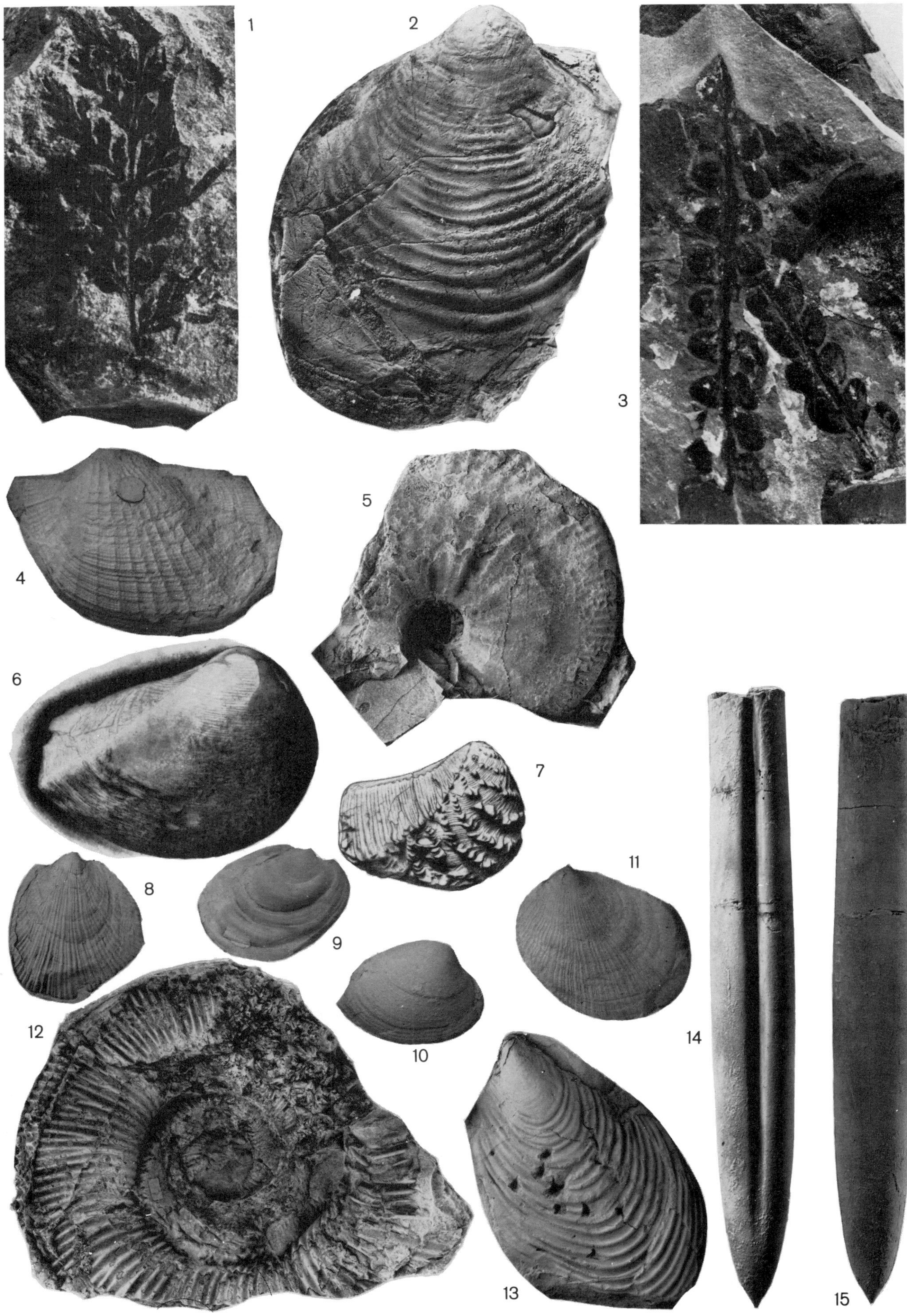
1
2
3
4
5
6
7
8
9
10
11
12
13
14
15

Plate 17

Representative Southwest Pacific Jurassic brachiopods and molluscs.

1, 2	***Pseudaucella marshalli*** (Trechmann). Ururoan Stage. 2 shows the characteristic crowded occurrence of this fossil. ×1.
3, 6	***Clavigera cuneiformis*** Hector. Aratauran Stage. ×1.
4, 7, 9–11	***Kalentera mackayi*** Marwick. Aratauran Stage. ×1.
5, 8	***Kalentera flemingi*** Marwick. Upper Ururoan Stage. ×1.
12, 14, 15	***Sphaeriola leedae*** Marwick. Aratauran Stage. ×1.5.
13	***Spiriferina radiata*** Hector. Ururoan Stage. ×2.
16	***Inoceramus*** cf. ***subhaasti*** Wandel. Heterian Stage. ×0.75.
17	***Inoceramus galoi*** Boehm. Heterian Stage. ×0.5.
18, 20	***Inoceramus haasti*** Zittel. Ohauan Stage. ×0.75.
19	***Spiriferina ongleyi*** Marwick. Aratauran Stage. ×1.

NZGS photos: D. L. Homer.
R. C. Brazier del.
Compiled: G. R. Stevens.

1
2
3
4
5
6
7
8
9
10
aa
pr
11
12
13
14
15
16
17
18
19
20

Plate 18

Some fossils from the Torlesse Supergroup (Triassic and Jurassic).

1 ***Titahia corrugata*** Webby. Late Triassic. ×0.5.

2, 3 ***Halorella ruahinensis*** Campbell. Otamitan–Warepan Stages. ×1.

4 ***Torlessia (=Terebellina) mackayi*** (Bather). ?Kaihikuan–Otamitan Stages. ×1.5.

5, 9 ***Holcothyris(?) kaiwaraensis*** Campbell. Late Jurassic. ×1.

6–8 ***Burmirhynchia warreni*** Campbell. Late Jurassic. ×1.

NZGS photos: L. R. Laronde (1) & D. L. Homer (4).
Otago University photos: R. H. Hardie (2, 3, 5–9).
Compiled: G. R. Stevens.

1
2
3
4
5
6
7
8
9

Plate 19

Representative Lower Cretaceous fossils.

1	***Inoceramus warakius*** Wellman. ?Urutawan–Motuan Stages. ×0.5.
2–4	***Aucellina*** cf. ***radiatostriata*** Bonarelli. Korangan Stage. 2, ×0.75; 3, ×1; 4, ×0.8.
5	***Trigonia*** n.sp. cf. ***tenuis*** Kitchin. Korangan Stage. ×1.5.
6	***Maccoyella*** cf. ***reflecta*** (Moore) (= *M. magnata* Marwick). Korangan Stage. ×0.5.
7–10	***Aucellina*** cf. ***gryphaeoides*** (Sowerby). Urutawan Stage. 7, 9, 10 ×1.5; 8, ×1.
11	***Inoceramus kapuus*** Wellman. Urutawan Stage. ×1.
12–14	***Aucellina euglypha*** Woods. Motuan Stage. ×1.5.
15	***Inoceramus ipuanus*** Wellman. Motuan Stage. ×0.5.
16, 17	***Wellmanites zelandicus*** Wright. Motuan Stage. ×1.5.
18, 19	***Dimitobelus superstes*** (Hector). Urutawan–Mangaotanean Stages. ×0.5.

NZGS photos: D. L. Homer.
Compiled: I. G. Speden.

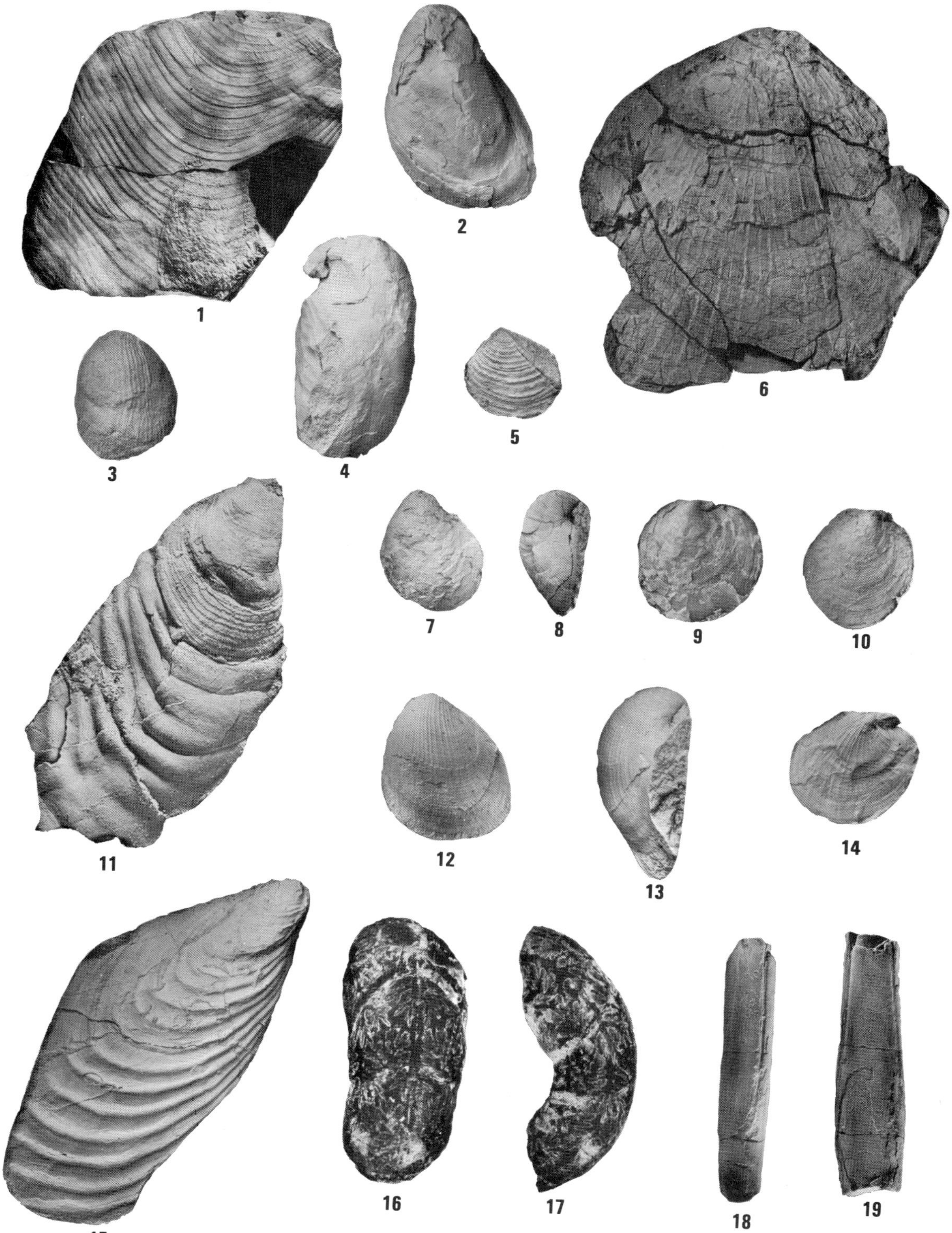

1
2
6
3
4
5
7
8
9
10
11
12
13
14
15
16
17
18
19

Plate 20

Representative Lower and Upper Cretaceous fossils.

1 ***Otoscaphites awanuiensis*** Wright. Ngaterian Stage. ×2.

2 ***Eselaevitrigonia meridiana*** (Woods). Ngaterian–Arowhanan Stages. ×1.

3, 4 ***Sciponoceras baculoide*** (Mantell). Ngaterian Stage. ×1.

5 ***Inoceramus hakarius*** Wellman. Ngaterian Stage. ×1.

6 ***Inoceramus fyfei*** Wellman. Ngaterian Stage. ×0.67.

7 ***Scaphites equalis coverhamensis*** Henderson. Ngaterian Stage. ×2.

8–10 ***Inoceramus tawhanus*** Wellman. Ngaterian Stage. ×0.5.

11, 12 ***Inoceramus concentricus*** Parkinson. Ngaterian Stage. ×0.5.

NZGS photos: D. L. Homer.
Compiled: I. G. Speden.

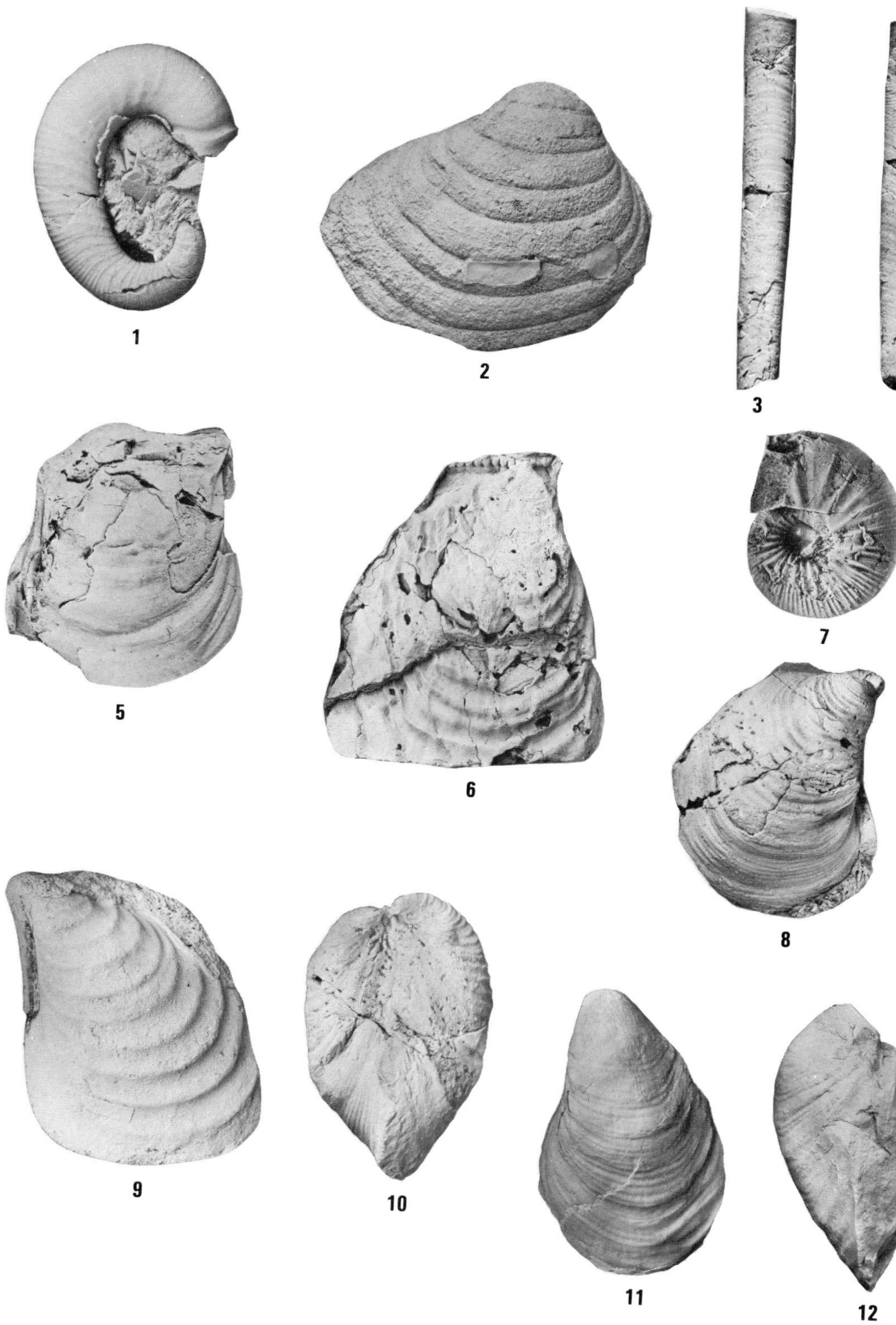
1
2
3
4
5
6
7
8
9
10
11
12

Plate 21

Representative Upper Cretaceous fossils.

1, 2 ***Inoceramus rangatira*** Wellman. Arowhanan Stage. ×0.125.

3 ***Iotrigonia glyptica*** (Woods). Ngaterian and Arowhanan Stages. ×0.75.

4, 5 ***Inoceramus bicorrugatus*** Marwick. Mangaotanean Stage. ×0.5.

6, 7 ***Inoceramus nukeus*** Wellman. Teratan Stage. 6, ×1; 7, ×0.8.

8, 9 ***Dimitobelus? ongleyi*** Stevens. Teratan–Haumurian Stages. ×0.75.

10 ***Inoceramus opetius*** Wellman. Teratan Stage. ×0.5.

NZGS photos: D. L. Homer.
Compiled: I. G. Speden.

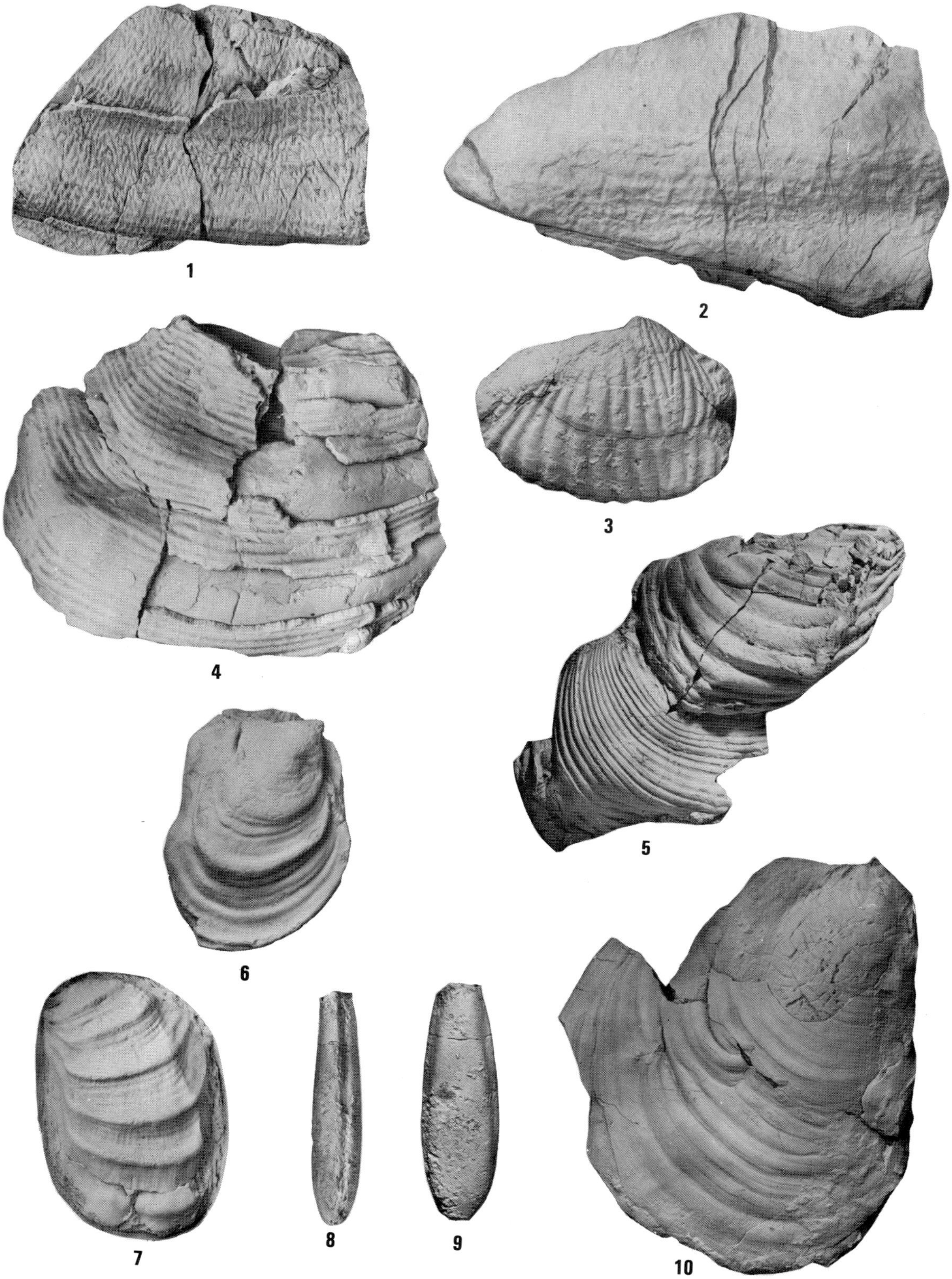
1
2
3
4
5
6
7
8
9
10

Plate 22

Representative Upper Cretaceous fossils.

1 ***Inoceramus pacificus*** Woods. Piripauan Stage. ×0.5.

2 ***Inoceramus australis*** Woods. Piripauan Stage. ×0.5.

3 ***Pterotrigonia pseudocaudata*** Hector. Piripauan and Haumurian Stages. ×0.75.

4 ***Kossmaticeras (K.) haumuriensis*** (Hector). Piripauan Stage. ×1.

5, 6 ***Dimitobelus lindsayi*** (Hector). Piripauan Stage. ×0.75.

7 ***Inoceramus matotorus*** Wellman. Haumurian Stage. ×0.25.

8 ***Ostrea lapillicola*** Marwick. Haumurian Stage, and possibly also Piripauan and Teratan Stages. ×2.

9 ***Perissoptera waiparaensis*** (Hector). Haumurian Stage. ×0.75.

10, 11 ***Dimitobelus hectori*** Stevens. Haumurian Stage. ×1.

NZGS photos: D. L. Homer
Compiled: I. G. Speden.

1

2

3

4

5

6

7

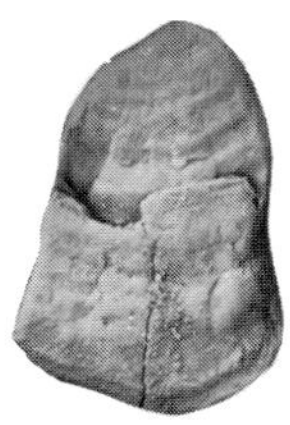

8

9

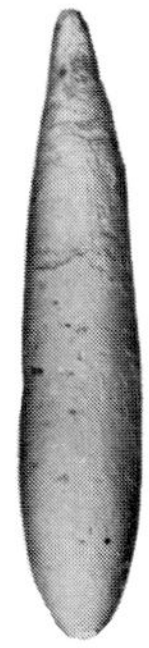

10

11

Plate 23

Representative Upper Cretaceous fossils.

1 ***Conchothyra parasitica*** (Hutton). Piripauan–Haumurian Stages. ×1.

2 ***Eriphyla meridiana*** (Woods). Piripauan–Haumurian Stages. ×1.

3 ***Lucina canterburiensis*** Woods. Piripauan–Haumurian Stages. ×1.

4 ***Lahillia (Lahilleona)*** cf. ***neozelanica*** Finlay & Marwick. Piripauan–Haumurian Stages. ×0.5.

5 ***Maorites tenuicostatus*** (Marshall). Piripauan–Haumurian Stages. ×0.75.

6 ***Vertebrites murdochi*** Marshall. Piripauan–Haumurian Stages. ×1.

7 ***Kossmaticeras (Natalites) sulcatum*** (Marshall). Piripauan–Haumurian Stages. ×0.9.

8 ***Gunnarites zelandicus*** (Marshall). Piripauan–Haumurian Stages. ×0.5.

9 ***Kossmaticeras (Natalites) bensoni*** Henderson. Piripauan–Haumurian Stages. ×0.25.

10, 11 ***Baculites rectus*** Marshall. Piripauan–Haumurian Stages. ×1.3.

NZGS photos: D. L. Homer.
Compiled: I. G. Speden.

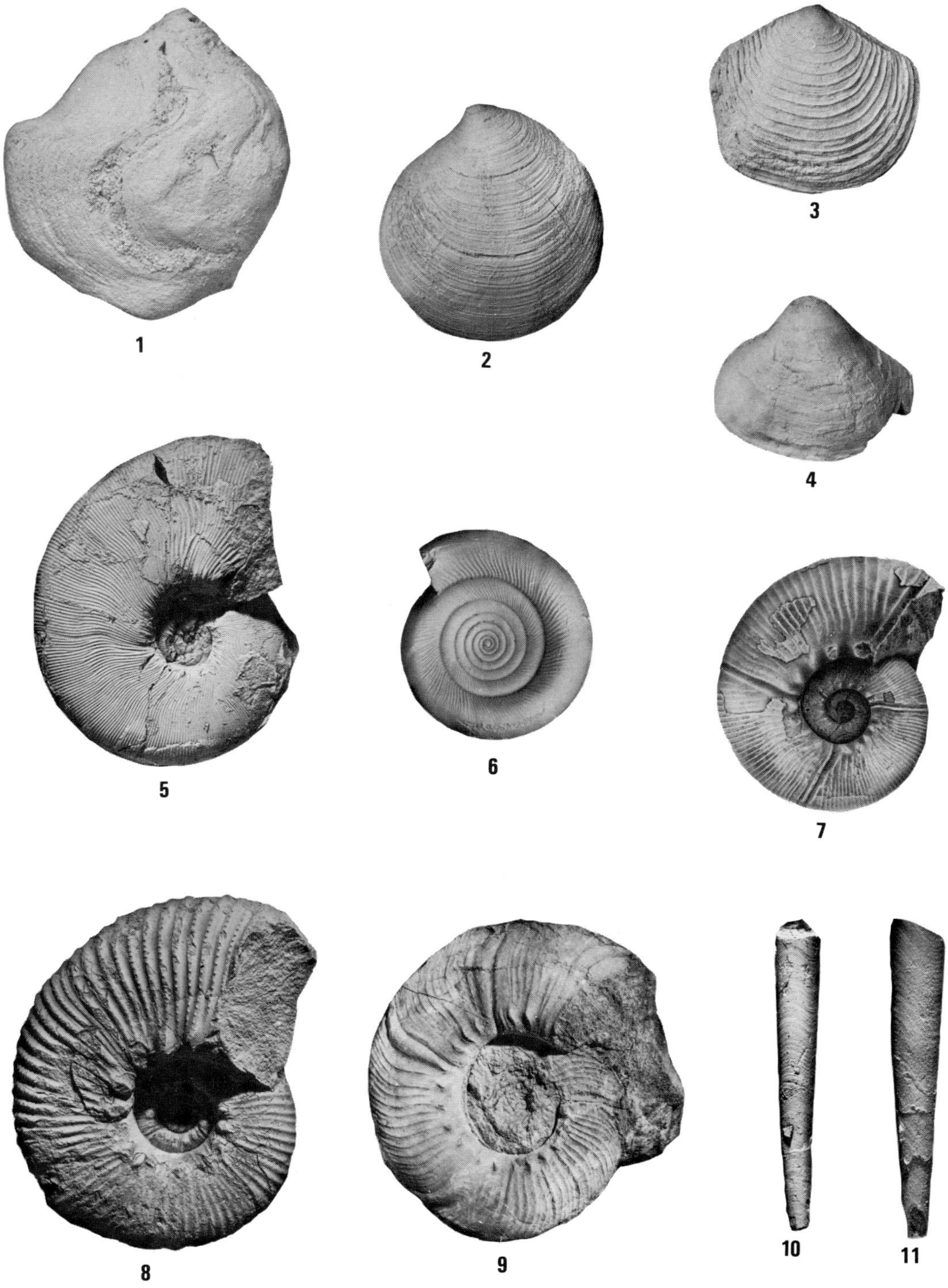
1
2
3
4
5
6
7
8
9
10
11

Plate 24

Representative Paleocene (Wangaloan) molluscs.

1 ***Spineilo elongata*** (Marshall). ×2.

2 ***Lahillia (Lahilleona) neozelanica*** Marshall & Murdoch. ×1.5.

3 ***Cucullaea (Cucullastis) barbara*** Finlay & Marwick. ×1.

4 ***Heteroterma zelandica*** Marshall. ×1.5.

5 ***Zeacolpus (Leptocolpus) semiconcavus*** (Suter). ×1.5.

6 ***Kaitangata hendersoni*** Finlay & Marwick. ×2.5.

7 ***Struthioptera osiris*** Finlay & Marwick. ×1.

8 ***Conchothyra australis*** (Marshall). ×1.5.

9 ***Ongleya tholispira*** Finlay & Marwick. ×3.

10 ***Campylacrum sanum*** Finlay & Marwick. ×6.5.

Drawings: Dr J. Marwick, originally published in
N.Z. Geological Survey Paleontological Bulletin 15.
Compiled: A. G. Beu & P. A. Maxwell.

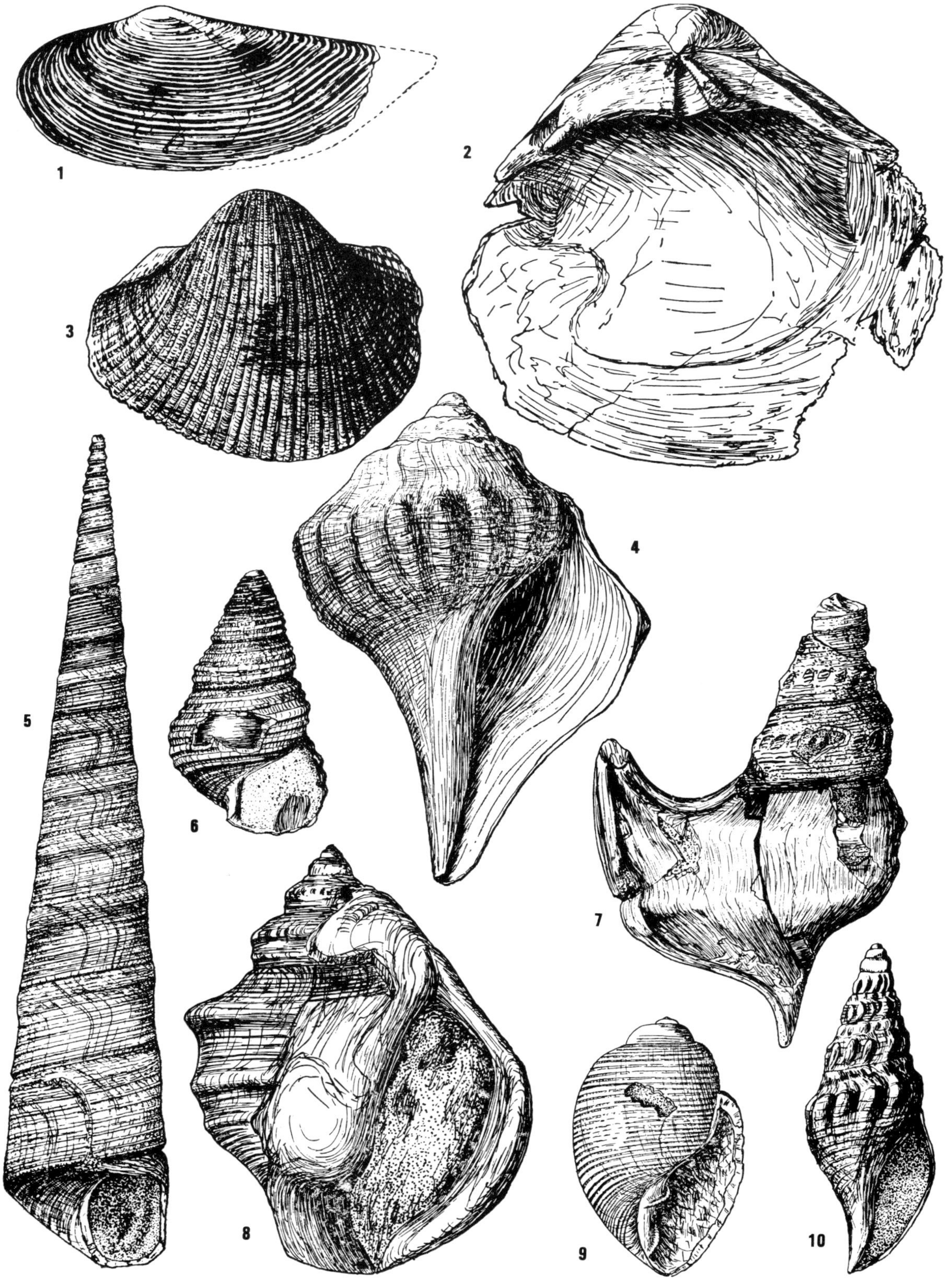
1
2
3
4
5
6
7
8
9
10

Plate 25

Representative Paleocene and Eocene fossils.

1 ***Serripecten venosus*** (Hutton). Runangan Stage. ×1.

2 ***Cubitostrea gudexi*** (Suter). Bortonian Stage. ×1.

3 ***Venericardia (Fasciculicardia) acanthodes*** (Suter). Bortonian Stage. ×1.

4 ***Pseudoportlandia solenelloides*** (Marshall). Bortonian Stage. ×1.

5 ***Notostrea tarda*** (Hutton). Waipawan Stage, Chatham Islands. ×1.

6 ***Notogenota finlayi*** Powell. Kaiatan Stage. ×1.

7 ***Carinacca waihaoensis*** (Suter). Kaiatan Stage. ×1.

8 ***Athleta necopinata necopinata*** (Suter). Bortonian Stage. ×1.

9 ***Monalaria concinna*** (Suter). Bortonian Stage. ×1.

10 ***Portunites granulifer*** Glaessner. Bortonian Stage. ×1.

11 ***Mauira curvispina*** (Marwick). Bortonian Stage. ×1.

12 ***Speightia spinosa*** (Suter). Bortonian Stage. ×1.

13 ***Fascioplex liraecostata*** (Suter). Bortonian Stage. ×1.

14 ***Colubraria neozelanica*** Maxwell. Kaiatan Stage. ×1.

NZGS photos: D. L. Homer & S. N. Beatus.
Compiled: A. G. Beu, I. W. Keyes, & P. A. Maxwell.

1
2
3
4
5
6
7
8
9
10
11
12
13
14

Plate 26

Representative Oligocene fossils.

1 ***Venericardia (Megacardita) ponderosa*** (Suter). Duntroonian Stage. ×0.65.

2 ***Lima colorata paleata*** Hutton. Whaingaroan–Waitakian Stages. ×0.75.

3 ***Serripecten (Janupecten) uttleyi*** (Marwick). Whaingaroan Stage. ×0.5.

4 ***Notocorbula humerosa*** (Hutton). Waitakian Stage. ×1.5.

5 ***Spissatella subobesa*** (Marshall & Murdoch). Duntroonian–Waitakian Stages. ×1.

6 ***Cirsotrema caelicola*** Finlay. Waitakian–Otaian Stages. ×1.

7 ***Magnatica planispira*** (Suter). Duntroonian–Altonian Stages. ×1.

8 ***Palaeeudyptes*** sp. (penguin). Runangan–Duntroonian Stages. ×1.

9 ***Histocidaris mckayi*** Fell (echinoid). Duntroonian Stage. ×0.75.

10 ***Kekenodon onamata*** Hector (whale tooth). Duntroonian Stage. ×0.65.

NZGS photos: D. L. Homer.

Compiled: A. G. Beu, I. W. Keyes, & P. A. Maxwell.

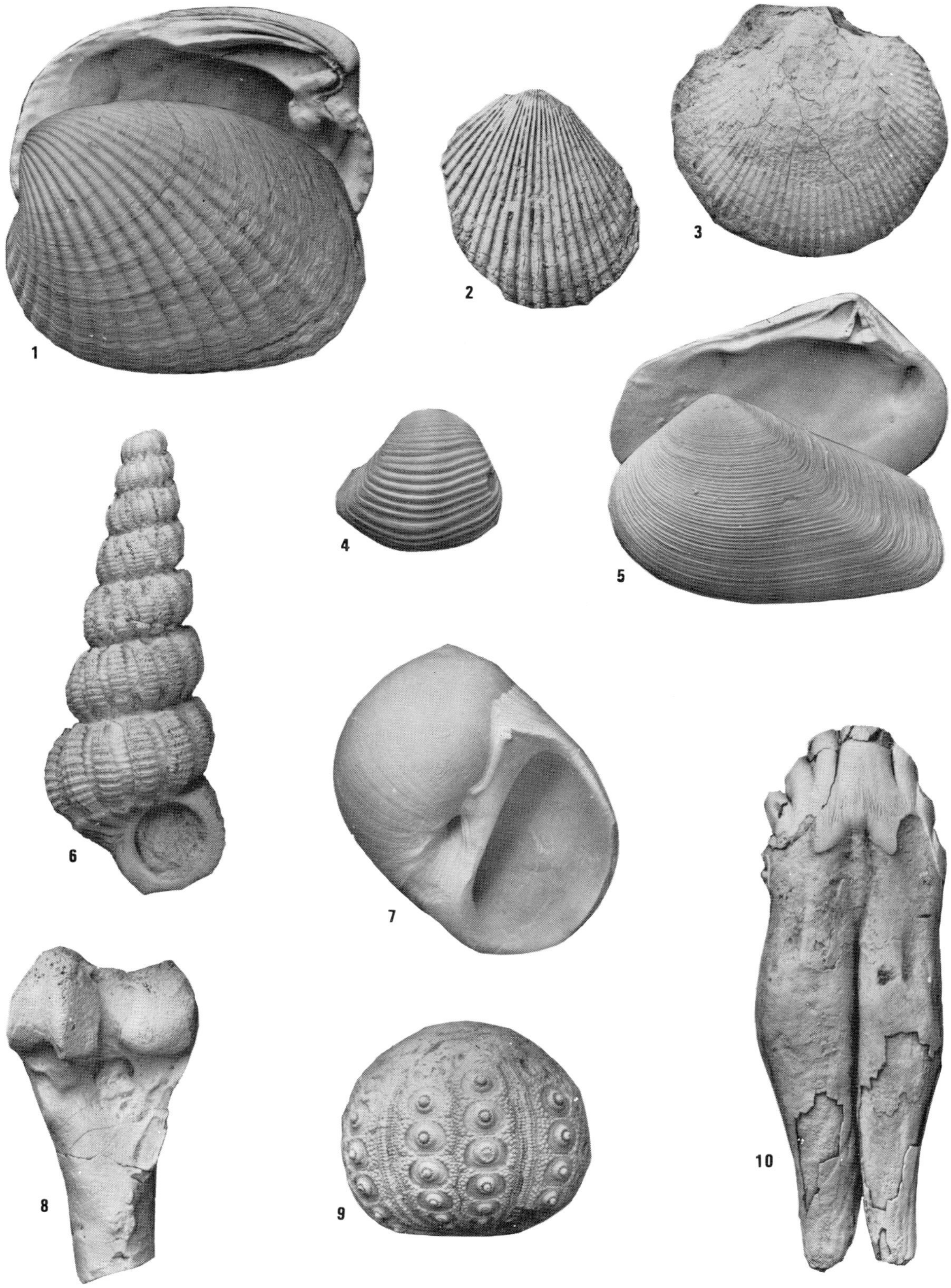
1
2
3
4
5
6
7
8
9
10

Plate 27

Representative Miocene bivalves.

1 ***Serripecten hutchinsoni*** (Hutton). Otaian–Opoitian Stages. ×0.65.

2 ***Bassina speighti*** (Suter). Duntroonian–Tongaporutuan Stages. ×1.

3 ***Venericardia (Fasciculicardia) subintermedia*** (Suter). Otaian–Lillburnian Stages. ×0.65.

4 ***Sectipecten wollastoni*** (Finlay). Kapitean Stage. ×0.65.

5 ***Lima colorata colorata*** Hutton. Otaian Stage – present day. ×0.65.

6 ***Lentipecten hochstetteri*** (Zittel). Duntroonian–Kapitean Stages. ×0.65.

7 ***Cucullaea (Latiarca) hamptoni*** Marwick. Tongaporutuan and Kapitean Stages. ×1.

8 ***Spissatella trailli*** (Hutton). Altonian Stage. ×1.

NZGS photos: D. L. Homer.
Compiled: A. G. Beu, I. W. Keyes, & P. A. Maxwell.

1
2
3
4
5
6
7
8

Plate 28

Representative Miocene gastropods.

1 ***Austrofusus (Austrofusus?) coerulescens*** (Finlay). Kapitean Stage. ×1.25.

2 ***Austrofusus (Neocola) apudalpha*** Finlay. Waitakian–Altonian Stages. ×1.

3 ***Austrofusus (Neocola) alpha*** Finlay. Early–mid Altonian Stage. ×1.

4 ***Austrofusus (Neocola) beta*** Finlay. Late Altonian Stage. ×1.

5 ***Austrofusus (Neocola) gamma*** Finlay. Clifdenian and Lillburnian Stages. ×1.

6 ***Austrofusus (Zelandiella) subnodosa*** (Hutton). Otaian–Waiauan Stages. ×1.

7 ***Austrofusus*** (s.str.) ***magnificus*** Finlay. Altonian–Lillburnian Stages. ×1.

8 ***Bathytoma (Bathytoma) bartrumi*** Laws. Otaian and Altonian Stages. ×1.

9 ***Austrotoma obsoleta*** Finlay. Waiauan and Tongaporutuan Stages. ×1.

10 ***Austrotoma echinata*** Powell. Altonian and Clifdenian Stages. ×1.

11 ***Maorivetia brevirostris*** (Hutton). Altonian and Clifdenian Stages. ×1.

12 ***Austrofusus (Zelandiella) fatua*** Finlay. Waiauan Stage. ×1.

13 ***Alcithoe (Mauithoe) insignis*** (Marwick). Lillburnian and Waiauan Stages. ×1.

14 ***Alcithoe (Waihaoia) phymatias*** (Finlay). Altonian–Lillburnian Stages. ×1.

15 ***Amalda (Baryspira) stortha*** (Olson). Altonian–Waiauan Stages. ×1.

16 ***Opella subfimbriata*** (Suter). Otaian and Altonian Stages. ×1.

17 ***Polinices (Polinices) huttoni*** (von Ihering). Otaian–Kapitean Stages. ×1.

18 ***Conus hendersoni*** Marwick. Tongaporutuan and Kapitean Stages. ×1.

19 ***Struthiolaria (Struthiolaria) subspinosa*** Marwick. Altonian–Waiauan Stages. ×1.

20 ***Struthiolaria (Callusaria) spinosa*** Hector. Altonian–Waiauan Stages. ×1.

21 ***Struthiolaria (Callusaria) callosa*** Marwick. Lillburnian–Tongaporutuan Stages. ×1.

NZGS photos: D. L. Homer & S. N. Beatus.
Compiled: A. G. Beu, I. W. Keyes, & P. A. Maxwell.

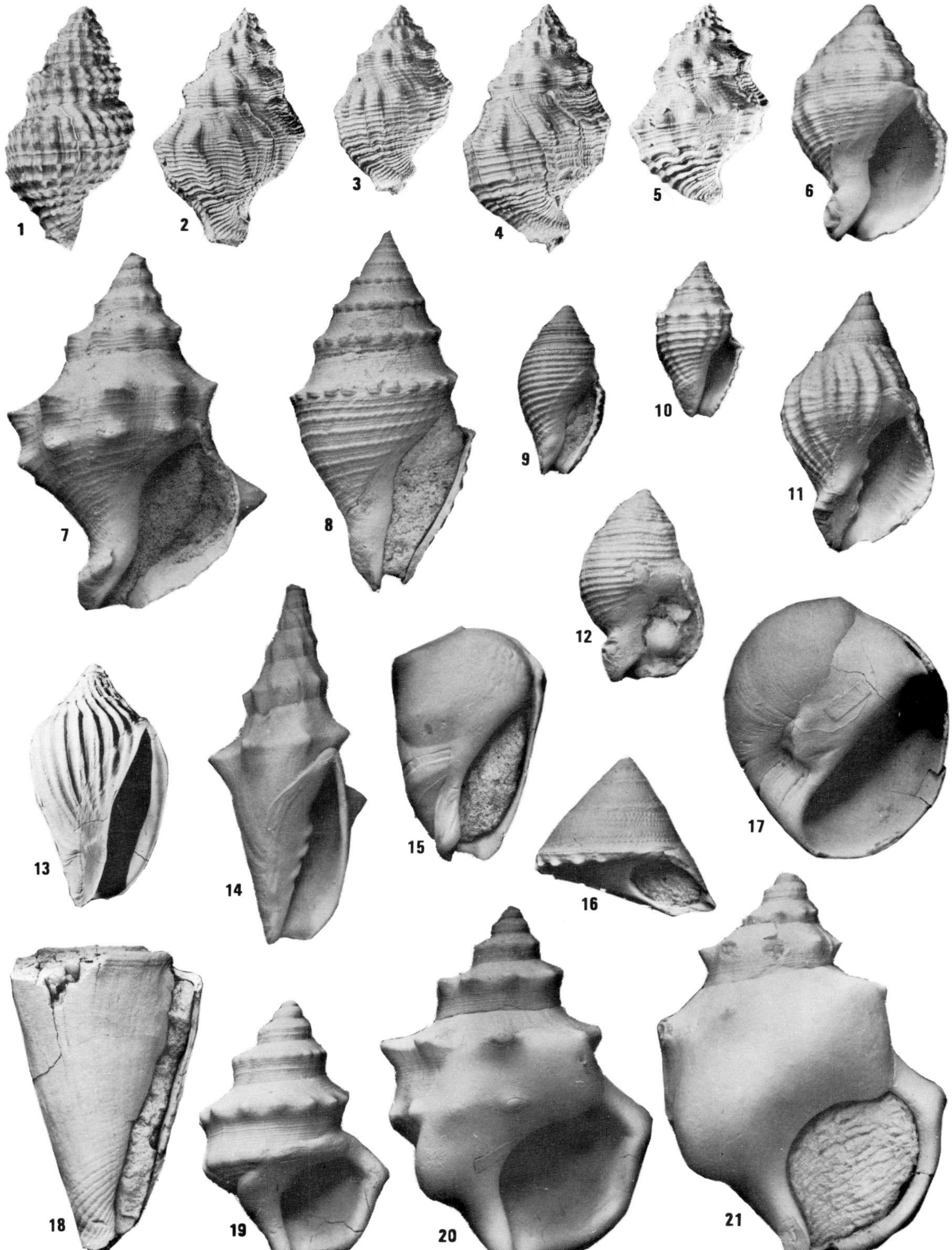
1
2
3
4
5
6
7
8
9
10
11
12
13
14
15
16
17
18
19
20
21

Plate 29

Representative Miocene molluscs with tropical affinities.

1 ***Chama huttoni*** Hector. Nukumaruan Stage (related forms Otaian–Opoitian Stages also). ×1.

2, 3 ***Spondylus aucklandicus*** Marshall. Otaian–Clifdenian Stages. ×1.

4 ***Tapes parki*** (Marshall & Murdoch). Duntroonian Stage. ×0.65.

5 ***Conilithes wollastoni*** Maxwell. Altonian Stage. ×1.

6 ***Cypraea (Notadusta) clifdenensis*** Cernohorsky. Otaian–Clifdenian Stages. ×2.25.

7 ***Chicoreus (Siratus) komiticus*** (Suter). Otaian and Altonian Stages. ×1.

8 ***"Marshallena" decens*** Marwick. Opoitian Stage. ×1.

9 ***Ficus altispira*** (Pritchard). Duntroonian–Otaian Stages. ×0.65.

10 ***Lyria zelandica*** Finlay. Altonian and Clifdenian Stages. ×1.

11 ***Morum (Oniscidea) finlayi*** (Laws). Altonian–Lillburnian Stages. ×1.

12 ***Echinophoria hectori*** (Abbott). Otaian and Altonian Stages. ×1.

NZGS photos: D. L. Homer.
Compiled: A. G. Beu.

1
2
3
4
5
6
7
8
9
10
11
12

Plate 30

Representative Pliocene molluscs.

1 ***Mesopeplum (Borehamia) crawfordi*** (Hutton). Waipipian Stage. ×0.65.

2 ***Chlamys (Phialopecten) triphooki triphooki*** (Zittel). Mangapanian and early Nukumaruan Stages. ×0.75.

3 ***Lima waipipiensis*** Marshall & Murdoch. Waipipian Stage. ×0.65.

4 ***Trachycardium (Ovicardium) rossi*** Marwick. Opoitian–Mangapanian Stages. ×0.65.

5 ***Maorimactra chrydaea*** (Suter). Kapitean and Waipipian Stage. ×1.5.

6 ***Hartungia typica typica*** Bronn. Opoitian and Waipipian Stages. ×1.

7 ***Clavatoma pulchra*** Powell. Opoitian–Mangapanian Stages. ×1.

8 ***Austrofusus (Austrofusus) pagoda*** (Finlay). Waipipian Stage. ×1.

9 ***Polinices (Polinices) waipipiensis*** (Marwick). Opoitian–Mangapanian Stages. ×1.

10 ***Struthiolaria (Pelicaria) rugosa*** Marwick. Late Mangapanian and early Nukumaruan Stages. ×1.

11 ***Struthiolaria (Pelicaria) canaliculata*** (Zittel). Waipipian Stage. ×1.

12 ***Struthiolaria (Pelicaria) zelandiae*** (Marshall & Murdoch). Opoitian and Waipipian Stages. ×1.

NZGS photos: D. L. Homer.
Compiled: A. G. Beu, I. W. Keyes, & P. A. Maxwell.

1
2
3
4
5
6
7
8
9
10
11
12

Plate 31

Representative Pleistocene molluscs.

1 ***Pecten benedictus marwicki*** (Finlay). Castlecliffian Stage. ×0.65.

2 ***Lutraria solida*** (Hutton). Waipipian–Nukumaruan Stages. ×0.65.

3 ***Barytellina tertia*** Fleming. Waipipian–Castlecliffian Stages. ×1.

4 ***Pecten novaezelandiae tainui*** Fleming. Castlecliffian Stage. ×0.65.

5 ***Protothaca (Tuangia)*** aff. ***crassicosta*** (Deshayes). Hawera Series. ×1.

6 ***Eumarcia plana*** Marwick. Waipipian–Nukumaruan Stages. ×0.65.

7 ***Glycymeris (Glycymeris) shrimptoni*** Marwick. Nukumaruan Stage. ×1.

8 ***Bassina parva parva*** Marwick. Nukumaruan Stage. ×1.

9 ***Struthiolaria (Pelicaria) convexa*** Marwick. Nukumaruan Stage. ×1.

10 ***Austrofusus (Austrofusus) glans glans*** (Röding). Castlecliffian Stage – present day. ×0.65.

11 ***Austrofusus (Austrofusus) glans chathamensis*** Finlay. Nukumaruan Stage – present day. ×0.65.

12 ***Leucotina ambigua*** (Hutton). Late Nukumaruan Stage – early Hawera Series. ×1.75.

13 ***Pterotyphis (Prototyphis) zelandicus*** (Hutton). Castlecliffian Stage – present day. ×1.

14 ***Struthiolaria (Pelicaria) vermis*** (Martyn). Castlecliffian – present day. ×1.

15 ***Struthiolaria (Pelicaria) acuminata*** Marwick. Late Mangapanian and early Nukumaruan Stages. ×1.

16 ***Struthiolaria (Struthiolaria) frazeri*** Hutton. Nukumaruan Stage. ×0.65.

NZGS photos: D. L. Homer.
Compiled: A. G. Beu.

1
2
3
4
5
6
7
8
9
10
11
12
13
14
15
16

Plate 32

Representative Pleistocene warm- and cold-water fossils.

1 ***Isognomon zealandicus*** (Suter). Kapitean–Nukumaruan Stages. ×0.5.

2, 3 ***Chlamys (Zygochlamys) delicatula*** (Hutton). Late Mangapanian Stage (?), Nukumaruan Stage – present day. ×1.

4 ***Aulacomya ater maoriana*** Iredale. Waipipian Stage – present day. ×0.65.

5 ***Anadara trapezia*** (Deshayes). Castlecliffian Stage – Hawera Series. ×0.65.

6 ***Philippia lutea*** (Lamarck). Nukumaruan Stage – present day. ×1.

7 ***Capulus uncinatus*** Hutton. Late Castlecliffian Stage. ×1.

8 ***Pterotyphis (Prototyphis) angasi powelli*** (Fleming). Castlecliffian Stage. ×1.

9 ***Balanus (Megabalanus) tubulatus*** Withers. Mangapanian and Nukumaruan Stages. ×0.65.

10 ***Semicassis pyrum*** (Lamarck). Nukumaruan Stage – present day. ×0.65.

11 ***Eunaticina cincta*** (Hutton). Late Castlecliffian Stage. ×1.5.

12 ***Argobuccinum tumidum*** (Dunker). Hawera Series and present day. ×0.65.

13 ***Calliostoma spectabile couperi*** (Vella). Nukumaruan Stage. ×1.

NZGS photos: D. L. Homer.
Compiled: A. G. Beu.

1
2
3
4
5
6
7
8
9
10
11
12
13

References Cited

Including references in Appendix

ADKIN, G.L.; COLLINS, B.W. 1967: A bibliography of New Zealand Geology to 1950. *N.Z. Geological Survey Bulletin 65.* 244 p.

ALLAN, R.S. 1935: The fauna of the Reefton beds, (Devonian) New Zealand. *N.Z. Geological Survey Paleontological Bulletin 14.* 72 p., pls 1-5.

——— 1947: A revision of the Brachiopoda of the Lower Devonian strata of Reefton, New Zealand. *Journal of Paleontology 21(5):* 436-52.

ARBER, E.A.N. 1917: The earlier Mesozoic floras of New Zealand. *N.Z. Geological Survey Paleontological Bulletin 6.* 80 p., pls 1-14.

ARMSTRONG, R.L.; McDOWALL, W.G. 1975: Proposed refinement of the Phanerozoic time scale. *Geodynamic Highlights 2:* 33-4.

BERGGREN, W.A. 1972: A Cenozoic time-scale — some implications for regional geology and paleobiogeography. *Lethaia 5:* 195-215.

BOREHAM, A.U.E. 1965: A revision of F.W. Hutton's pelecypod species described in the Catalogue of Tertiary Mollusca and Echinodermata (1873). *N.Z. Geological Survey Paleontological Bulletin 37.* 125 p., pls 1-20.

BOREHAM, A.U.E.; FLEMING, C.A. 1957: Notes on the paleontology of the concretions. Appendix pp. 788-91 *in* PANTIN, H.M., Fossiliferous concretions from the shelf south-east of Cape Campbell, New Zealand. *N.Z. Journal of Science & Technology B, 38(7):* 781-91.

BOUCOT, A.J.; CASTER, K.E.; IVES, D.; TALENT, J.A. 1963: Relationships of a new Lower Devonian terebratuloid (Brachiopoda) from Antarctica. *Bulletins of American Paleontology 46(207):* 81-151, pls 16-41.

BROWNE, R.A.S. 1952: Early Triassic Ammonoids from Beaumont Station, Wairaki Survey District. *Transactions of the Royal Society of N.Z. 79(3&4):* 528-34.

CAMPBELL, J.D. 1965: New species of brachiopods from the Torlesse Group of Kaiwara Valley, north Canterbury. *Transactions of the Royal Society of N.Z., Geology 3(7):* 95-7.

——— 1968: *Rastelligera* (Brachiopoda) of the Upper Triassic of New Zealand. *Transactions of the Royal Society of N.Z., Geology 6(3):* 23-37.

CLARKE, M.J. 1979: The Tasmanian Permian spiriferid brachiopods *Trigonotreta stokesi*, Koenig, 1825, *Grantonia hobartensis*, Brown, 1953 and *Spirifer tasmaniensis*, Morris, 1845. *Journal of Paleontology 53(1):* 197-207.

COOPER, R.A. 1979: Ordovician geology and graptolite faunas of the Aorangi Mine area, North-west Nelson, New Zealand. *N.Z. Geological Survey Paleontological Bulletin 47.* 121 p., pls 1-19.

DAY, R.W.; SPEDEN, I.G. 1970: Probable synonymy of the Lower Cretaceous bivalves *Maccoyella reflecta* (Moore) and *Maccoyella magnata* Marwick. *N.Z. Journal of Geology & Geophysics 13(3):* 647-54.

FELL, H.B. 1954: Tertiary and Recent Echinoidea of New Zealand: Cidaridae. *N.Z. Geological Survey Paleontological Bulletin 23.* 62 p., pls 1-15.

FINLAY, H.J.; MARWICK, J. 1937: The Wangaloan and associated molluscan faunas of Kaitangata - Green Island Subdivision. *N.Z. Geological Survey Paleontological Bulletin 15.* 140 p., pls 1-18.

FLEMING, C.A. 1951: The genus *Pecten* in the west Pacific. *Journal de Conchyliologie 90:* 276-82.

——— 1957a: Lower Devonian Pelecypoda from Reefton, New Zealand. *Transactions of the Royal Society of N.Z. 85 (1):* 135-40.

——— 1957b: The genus *Pecten* in New Zealand. *N.Z. Geological Survey Paleontological Bulletin 26.* 69 p., pls 1-15.

——— 1958: Upper Jurassic fossils and hydrocarbon traces from the Cheviot Hills, north Canterbury. *N.Z. Journal of Geology & Geophysics 1(2):* 375-94.

——— 1959: *Buchia plicata* (Zittel) and its allies, with a description of a new species, *Buchia hochstetteri*. *N.Z. Journal of Geology & Geophysics 2(5):* 889-904.

——— 1962a: Palaeontological evidence for speciation preceded by geographic isolation. Pp. 225-41, (6 figs, pls 1-3) *in* LEEPER, G.W. (Ed.) The Evolution of Living Organisms - Symposium Volume, Royal Society of Victoria. Melbourne Univesity Press. 459 p.

——— 1962b: The genus *Pterynotus* Swainson (Gastropoda, Family Muricidae) in New Zealand and Norfolk

Island. *Transactions of the Royal Society of N.Z., Zoology 2(14):* 109-19.

FLEMING, C.A. 1962c: Two new genera of Triassic Trigoniidae from New Zealand. *Proceedings of the Malacological Society of London 35(1):* 1-4.

——— 1963: A new genus and species of Trigoniidae from the Middle Triassic of New Zealand. *N.Z. Journal of Geology & Geophysics 6(5):* 843-6.

——— 1964: History of the Bivalve Family Trigoniidae in the South-west Pacific. *The Australian Journal of Science 26(7):* 196-204.

——— 1966: Marwick's illustrations of New Zealand shells, with a checklist of New Zealand Cenozoic Mollusca. *N.Z. Department of Scientific & Industrial Research Bulletin 173.* 456 p., pls 1-145.

——— 1970: The Mesozoic of New Zealand: chapters in the history of the circum-Pacific mobile belt. *Quarterly Journal of the Geological Society of London 125:* 125-70.

GAIR, H.S.; GREGG, D.R.; SPEDEN, I.G. 1962: Triassic fossils from Corbies Creek, North Otago. *N.Z. Journal of Geology & Geophysics 5(1):* 92-113.

GLAESSNER, M.F. 1960: The fossil Decapod Crustacea of New Zealand and the evolution of the Order Decapoda. *N.Z. Geological Survey Paleontological Bulletin 31.* 79 p., pls 1-7.

GRANT-MACKIE, J.A. 1960: On a new *Kalentera* (Pelecypoda: Cypricardiacea) from the Upper Triassic of New Zealand. *N.Z. Journal of Geology & Geophysics 3(1):* 74-80.

——— 1976: The Upper Triassic bivalve *Monotis* in the southwest Pacific. *Pacific Geology 11:* 47-56.

GRINDLEY, G.W.; HARRINGTON, H.J.; WOOD, B.L. 1959: The geological map of New Zealand 1:2 000 000. *N.Z. Geological Survey Bulletin 66.* 111 p., pls 1-5.

HARDENBOL, J.Q.; BERGGREN, W.A. 1978: Contributions to the geologic time scale. *American Association of Petroleum Geologists, Studies in Geology 6:* 213-34.

HENDERSON, R.A. 1970: Ammonoidea from the Mata Series (Santonian - Maastrichtian) of New Zealand. *The Palaeontological Association, London: Special Papers in Palaeontology No. 6.* 82 p

——— 1973: Clarence and Raukumara Series (Albian - ?Santonian) ammonoidea from New Zealand. *Journal of the Royal Society of N.Z. 3(1):* 71-123.

HILL, D. 1956: Paleozoic corals from New Zealand. Part 1. The Devonian corals of Reefton, New Zealand. *N.Z. Geological Survey Paleontological Bulletin 25:* 1-14, pls 1, 2.

HORNIBROOK, N.deB. 1968: A handbook of New Zealand microfossils (Foraminifera and Ostracoda). *N.Z. Department of Scientific and Industrial Research, Information Series 62.* 135 p.

——— 1977: The Neogene (Miocene - Pliocene) of New Zealand. *Proceedings First International Congress on Pacific Neogene Stratigraphy, Tokyo 1976:* 145-50.

JENKINS, D.L. 1976: Index to Bibliography of New Zealand Geology to 1950 (N.Z. Geological Survey Bulletin 65). *N.Z. Geological Survey Bulletin 65A.* 242 p.

KUMMEL, B. 1959: Lower Triassic Ammonoids from western Southland, New Zealand. *N.Z. Journal of Geology & Geophysics 2(3):* 429-47.

——— 1960: New Zealand Triassic ammonoids. *N.Z. Journal of Geology & Geophysics 3(3):* 486-509.

——— 1965: New Lower Triassic ammonoids from New Zealand. *N.Z. Journal of Geology & Geophysics 8(3):* 537-47.

LEED, H. 1956: Paleozoic corals from New Zealand. Part 2. Permian reef-building corals from North Auckland Peninsula, New Zealand. *N.Z. Geological Survey Paleontological Bulletin 25:* 15-23, pls 3-5.

MARPLES, B.J.; FLEMING, C.A. 1963: A fossil penguin bone from Kawhia, New Zealand. *N.Z. Journal of Geology & Geophysics 6:* 189-92.

MARSHALL, P. 1917: The Wangaloa beds. *Transactions of the N.Z. Institute 49:* 450-60, pls 34-37.

——— 1926: The Upper Cretaceous ammonites of New Zealand. *Transactions of the N.Z. Institute 56:* 129-210.

MARSHALL, P.; MURDOCH, R. 1919: Some new fossil species of Mollusca. *Transactions of the N.Z. Institute 51:* 253-8.

MARWICK, J. 1923: The genus *Glycymeris* in the Tertiary of New Zealand. *Transactions of the N.Z. Institute 54:* 63-80.

——— 1924a: An examination of some of the Tertiary Mollusca claimed to be common to Australia and New Zealand. *Report of the Australasian Association for the Advancement of Science 16:* 316-31, pls 5-6.

MARWICK, J. 1924b: The Struthiolariidae. *Transactions of the N.Z. Institute 55:* 161-90.

———— 1924c: The Tertiary and Recent Naticidae and Naricidae of New Zealand. *Transactions of the N.Z. Institute 55:* 545-79.

———— 1926a: Tertiary and Recent Volutidae of New Zealand. *Transactions and Proceedings of the N.Z. Institute 56:* 259-303, pls 61-70.

———— 1926b: Cretaceous fossils from Waiapu Subdivision. *N.Z. Journal of Science & Technology 8(6):* 379-82.

———— 1927: The Veneridae of New Zealand. *Transactions of the N.Z. Institute 57:* 567-635.

———— 1928: The Tertiary Mollusca of the Chatham Islands including a generic revision of the New Zealand Pectinidae. *Transactions of the N.Z. Institute 58:* 432-506.

———— 1931: The Tertiary Mollusca of the Gisborne District. *N.Z. Geological Survey Paleontological Bulletin 13.* 177 p., pls 1-18.

———— 1939: *Maccoyella* and *Aucellina* in the Taitai Series. *Transactions & Proceedings of the Royal Society of N.Z. 68:* 462-5.

———— 1943: Some Tertiary Mollusca from North Otago. *Transactions of the Royal Society of N.Z. 73:* 181-92, pls 25-27.

———— 1944: New Zealand fossil and Recent Cardiidae (Mollusca). *Transactions of the Royal Society of N.Z 74:* 255-72.

———— 1948: Lower Pliocene Mollusca from Otahuhu, Auckland. *N.Z. Geological Survey Paleontological Bulletin 16.* 38 p., pls 1-8.

———— 1953: Divisions and faunas of the Hokonui System (Triassic and Jurassic). *N.Z. Geological Survey Paleontological Bulletin 21.* 141 p., pls 1-17.

———— 1965: Upper Cenozoic Mollusca of Wairoa District, Hawke's Bay. *N.Z. Geological Survey Paleontological Bulletin 39.* 83 p., pls 1-11.

———— 1966: An aberrant aucellinoid (Bivalvia, Pteriacea) from Red Island, Hawke's Bay. *N.Z. Journal of Geology & Geophysics 9(4):* 495-503.

———— 1971: New Zealand Turritellidae related to *Zeacolpus* Finlay (Gastropoda). *N.Z. Geological Survey Paleontological Bulletin 44.* 87 p., pls 1-10.

MAXWELL, P.A. 1966: Some upper Eocene Mollusca from New Zealand. *N.Z. Journal of Geology & Geophysics 9(4):* 439-57.

———— 1969: Middle Tertiary Mollusca from north Otago and south Canterbury, New Zealand. *Transactions of the Royal Society of N.Z., Geology 6(13):* 155-85.

MILNE, J.D.G.; CAMPBELL, J.D. 1969: Upper Triassic fossils from Oroua Valley, Ruahine Range, New Zealand. *Transactions of the Royal Society of N.Z., Geology 6(18):* 247-50.

ODIN, G.S. 1978: Results of dating Cretaceous - Paleogene sediments, Europe. Contributions to the geologic time scale. *American Association of Petroleum Geologists, Studies in Geology 6:* 127-41.

OLSON, O.P. 1956: The genus *Baryspira* (Mollusca) in New Zealand. *N.Z. Geological Survey Paleontological Bulletin 24.* 32 p., Pls 1-7.

POWELL, A.W.B. 1942: The New Zealand Recent and fossil Mollusca of the Family Turridae. *Auckland Institute & Museum Bulletin 2.* 188 p., pls 1-14.

REED, F.R.C. 1927: New trilobites from the Ordovician beds of New Zealand. *Transactions of the N.Z. Institute 57:* 310-4.

SHIRLEY, J. 1938: The fauna of the Baton River beds (Devonian), New Zealand. *Quarterly Journal of the Geological Society of London 94(4):* 459-506.

SKWARKO, S.K. 1962: Graptolites of Cobb River - Mount Arthur area, north-west Nelson, New Zealand. *Transactions of the Royal Society of N.Z., Geology 1(15):* 215-47.

SPEDEN, I.G. 1970: Three new inoceramid species from the Jurassic of New Zealand. *N.Z. Journal of Geology & Geophysics 13(3):* 825-51.

STEVENS, G.R. 1965: The Jurassic and Cretaceous belemnites of New Zealand and a review of the Jurassic and Cretaceous belemnites of the Indo-Pacific region. *N.Z. Geological Survey Paleontological Bulletin 36.* 283 p., pl. 1-25.

STEVENS, G.R.; SPEDEN, I.G. 1978: New Zealand. Pp. 251-328 *in* MOULLADE, M.; NAIRN, A.E.M. (Eds). The Phanerozoic Geology of the World, II: The Mesozoic, A. Elsevier Scientific Publishing Co., Amsterdam:

SUGGATE, R.P.; RIDDOLLS, P.M. 1976: Geology. Pp. 90-97 *in* WARDS, I.McL. (Ed.). New Zealand Atlas. Government Printer, Wellington.

SUGGATE, R.P.; STEVENS, G.R.; TE PUNGA, M.T. (Eds) 1979: The Geology of New Zealand. Government Printer. Wellington, 2 vols, 820 p.

SUTER, H. 1914: Revision of the Tertiary Mollusca of New Zealand, based on type material. Part I. *N.Z. Geological Survey Paleontological Bulletin 2.* 64 p., pls 1-17.

——— 1915: Revision of the Tertiary Mollusca of New Zealand, based on type material. Part II. *N.Z. Geological Survey Paleontological Bulletin 3.* 69 p., pls 1-10.

TRECHMANN, C.T. 1918: The Trias of New Zealand. *Quarterly Journal of the Geological Society of London 73:* 165-246.

VAN HINTE, J.E. 1976: A Jurassic time scale. *Bulletin American Association of Petroleum Geologists 60(4):* 489-97.

VELLA, P. 1954: Tertiary Mollusca from south-east Wairarapa. *Transactions of the Royal Society of N.Z. 81(4):* 539-55.

WARREN, Guyon; SPEDEN, Ian; HODDER, Peter; SEARLE, E.J. 1977: A bibliography of New Zealand Geology 1951-1969. *N.Z. Geological Survey Bulletin 93,* 193 p.

WATERHOUSE, J.B. 1960a: *Mellarium,* a new Triassic pleurotomariid gastropod from New Zealand. *N.Z. Journal of Geology & Geophysics 3(2):* 271-83.

——— 1960b: Some Carnian pelecypods from New Zealand. New Zealand. *Transactions of the Royal Society of N.Z. 88(3):* 425-42.

——— 1961: A new species of *Worthenia* (Gastropoda) from the Triassic of Southland, New Zealand. *Transactions of the Royal Society of N.Z., Geology 1(2):* 5-9.

——— 1963a: Permian Gastropods of New Zealand. Part 1 - Bellerophontacea and Euomphalacea. *N.Z. Journal of Geology & Geophysics 6(1):* 88-112.

——— 1963b: Permian gastropods of New Zealand. Part 2- Pleurotomariacea (in part). *N.Z. Journal of Geology & Geophysics 6(2):* 115-54.

——— 1963c: Permian gastropods of New Zealand. Part 3 - Pleurotomariacea (concluded). *N.Z. Journal of Geology & Geophysics 6(4):* 587-622.

——— 1963d: New Zealand species of the Permian bivalve *Atomodesma* Beyrich. *Palaeontology 6(4):* 699-717.

——— 1963e: *Etheripecten,* a new aviculopectinid genus from the Permian. *N.Z. Journal of Geology & Geophysics 6(2):* 193-6.

——— 1964a: Permian brachiopods of New Zealand. *N.Z. Geological Survey Paleontological Bulletin 35.* 212 p., pls 1-37.

——— 1964b: A Permian ammonoid from New Zealand. *Journal of Paleontology 38(1):* 149-50.

——— 1966: The age of the Croisilles Volcanics, eastern Nelson. *Transactions of the Royal Society of N.Z., Geology 3(13):* 175-81.

——— 1967: Upper Permian (Tatarian) brachiopods from New Zealand. *N.Z. Journal of Geology & Geophysics 10(1):* 74-118.

——— 1968: The classification and descriptions of Permian Spiriferida (Brachiopoda) from New Zealand. *Palaeontographica 129A.* 94 p., pls 1-18.

——— 1969: The Permian bivalve genera *Myonia, Megadesmus, Vacunella* and their allies, and their occurrences in New Zealand. *N.Z. Geological Survey Paleontological Bulletin 41.* 141 p., pls 1-23.

——— 1973: An ophiceratid ammonoid from the New Zealand Permian and its implications for the Permian Triassic boundary. *Geological Magazine 110(4):* 305-84.

——— 1976: World correlations for marine Permian faunas. *Queensland University Department of Geology Papers, 7(2):* I-XVII, 1-232.

——— 1978: Chronstratigraphy for the World Permian. Contributions to the Geologic Time Scale. *American Association of Petroleum Geologists, Studies in Geology 6:* 299-322.

——— 1979: New members of the Atomodesminae (Bivalvia) from the Permian of Australia and New Zealand. *Papers Department of Geology University of Queensland 9(1):* 1-22.

WATERHOUSE, J.B.; VELLA, P. 1965: A Permian fauna from north-west Nelson, New Zealand. *Transactions of the Royal Society of N.Z., Geology 3(5):* 57-84.

WEBBY, B.D. 1958: A lower Mesozoic annelid from Rock Point, south-western Wellington, New Zealand. *N.Z. Journal of Geology & Geophysics 1(3):* 509-13.

WELLMAN, H.W. 1959: Divisions of the New Zealand Cretaceous. *Transactions of the Royal Society of N.Z. 87(1 & 2):* 99-163.

WILCKENS, O. 1922: The Upper Cretaceous gastropods of New Zealand. *N.Z. Geological Survey Paleontological Bulletin 9.* 42 p., pls 1-5.

WILCKENS, O. 1927: Contributions to the Palaeontology of the New Zealand Trias. *N.Z. Geological Survey Paleontological Bulletin 12*. 65 p., pls 1-10.

WITHERS, T.H. 1924: The fossil cirripedes of New Zealand. *N.Z. Geological Survey Paleontological Bulletin 10*. 47 p., pls 1-8.

WOODS, H. 1917: The Cretaceous faunas of the north-eastern part of the South Island of New Zealand. *N.Z. Geological Survey Paleontological Bulletin 4*. 41 p., pls 1-20.

WRIGHT, C.W. 1957: Some Cretaceous ammonites from New Zealand. *Transactions of the Royal Society of N.Z. 84(4)*: 805-9.

Further Reading

Articles giving information on New Zealand fossils and geology are scattered through scientific and popular literature. Those interested in obtaining further information should consult, particularly, issues of the N.Z. Journal of Geology and Geophysics (1958 -), the Journal of the Royal Society (1971 -), and their forerunners, the New Zealand Journal of Science and Technology (1918 - 1957) and the Transactions of the Royal Society (1869 - 1971). Paleontological bulletins of New Zealand Geological Survey are fundamental sources for descriptions and illustrations of major groups of fossils. Location of articles of interest can be facilitated by the use of indexes, usually annual, in issues of journals, and of bibliographies (Adkin & Collins 1967, Warren *et al.* 1977) and related indexes (Jenkins 1976).

Important references covering macrofossils include:

GENERAL

CAMPBELL, J.D. 1975: Classic fossil localities. *New Zealand's Nature Heritage 83:* 2325-8.

CARTER, R.M. 1974: Geographies of the past (1). *New Zealand's Nature Heritage 1(4):* 102-7.

———— 1974 Geographies of the past (2). *New Zealand's Nature Heritage 1(5):* 129-35.

FLEMING, C. A. 1979: The Geological history of New Zealand and its life. Auckland University Press and Oxford University Press, Auckland, 141 p.

HORNIBROOK, N. de B. 1968: A handbook of New Zealand microfossils (Foraminifera and Ostracoda). *N.Z. Department of Scientific and Industrial Research, Information Series 62,* 135 p.

MILDENHALL, D.M. 1974: The record of the rocks. *New Zealand's Nature Heritage 2:* 43-7.

STEVENS, G.R.; HORNIBROOK, N. de B.; HARRIS, W.F. 1966: Fossils. *In* McLintock, A.H. (Ed.). An Encyclopedia of New Zealand 1: 733-42.

PALEOZOIC

ALLAN, R.S. (1935): The Fauna of the Reefton Beds (Devonian) New Zealand; with notes on Lower Devonian Animal Communities in Relation to the Base of the Devonian System. *N.Z. Geological Survey Paleontological Bulletin 14.*

———— 1947: A revision of the Brachiopods of the Lower Devonian strata of Reefton, New Zealand. *Journal of Paleontology 21(5):* 436-52.

BENSON, W.N.; KEBLE, R.A. 1935: The Geology of the Regions Adjacent to Preservation and Chalkey Inlets, Fiordland, New Zealand. Part IV. Stratigraphy and Paleontology of the Fossiliferous Ordovician Rocks. *Transactions of the Royal Society of N.Z. 65:* 244-94.

KEBLE, R.A.; BENSON, W.N. (1929): Ordovician Graptolites of North-west Nelson. *Transactions of the N.Z. Institute 59:* 840-63.

SHIRLEY, J. (1938): The Fauna of the Baton River Beds (Devonian), New Zealand. *Quarterly Journal of the Geological Society of London 94:* 459-506.

SKWARKO, S.K. 1962: Graptolites of Cobb River - Mount Arthur Area, Northwest Nelson, New Zealand. *Transactions of the Royal Society of N.Z. 1(15):* 215-247.

WATERHOUSE, J.B. 1964: Permian brachiopods of New Zealand. *N.Z. Geological Survey Paleontological Bulletin 35:* 287 pp., 37 pls.

———— 1968: The classification and descriptions of Permian Spiriferida (Brachiopoda) from New Zealand. *Palaeontographica 129A:* 94 pp., 18 pls.

WATERHOUSE, J.B. 1969: The Permian bivalve genera *Myonia, Megadesmus, Vacunella* and their allies, and their occurrences in New Zealand. *N.Z. Geological Survey Paleontological Bulletin 41:* 141 p., 53 pls.

MESOZOIC

CAMPBELL, J.D. 1968: *Rastelligera* (Brachipoda) of the upper Triassic of New Zealand. *Transactions of the Royal Society of N.Z., Geology 6(3):* 23-37.

FLEMING, C.A. 1959: *Buchia plicata* (Zittel) and its allies, with a description of a new species, *Buchia hochstetteri. N.Z. Journal of Geology and Geophysics 2(5):* 889-904.

GRANT-MACKIE, J.A. 1978: Subgenera of the Upper Triassic bivalve *Monotis. N.Z. Journal of Geology and Geophysics 21(1):* 97-111.

(See also subsequent papers in the same journal and in the Journal of the Royal Society of New Zealand, 1978-9.)

HENDERSON, R.A. 1970: Ammonoidea from the Mata Series. (Santonian - Maastrichtian) of New Zealand. *Palaeontological Association of London, Special Papers on Palaeontology 6.* 82 p., 15 pls.

——————— 1973: Clarence and Raukumara Series (Albian - ?Santonian) Ammonoidea from New Zealand. *Journal of the Royal Society of N.Z. 3:* 71-123.

KUMMEL, B. 1960: New Zealand Triassic ammonoids. *N.Z. Journal of Geology and Geophysics 3(3):* 486-509.

MARWICK, J. 1953: Divisions and faunas of the Hokonui System (Triassic and Jurassic). *N.Z. Geological Survey Paleontological Bulletin 21:* 141 p.,17 pls.

STEVENS, G.R. 1966: The Jurassic and Cretaceous belemnites of New Zealand and a review of the Jurassic and Cretaceous belemnites of the Indo-Pacific Region. *N.Z. Geological Survey Paleontological Bulletin 36:* 283 p., 25 pls.

WARREN, Guyon; SPEDEN, Ian 1977: The Piripauan and Haumurian Stratotypes (Mata Series, Upper Cretaceous) and correlative sequences in the Haumuri Bluff District, South Marlborough. *N.Z. Geological Survey Bulletin 92:* 59 p. (Upper Cretaceous fossils illustrated on figures 25 and 26.)

WELLES, S.P.; GREGG, D.R. 1971: Late Cretaceous marine reptiles of New Zealand. *Records of the Canterbury Museum 9(1):* 1-111.

WILCKENS, O. 1922: The Upper Cretaceous Gastropods of New Zealand. *N.Z. Geological Survey Paleontological Bulletin Bulletin 9:* 42 p., 5 pls.

WOODS, H. 1917: The Cretaceous faunas of the north-eastern part of the South Island. *N.Z. Geological Survey Paleontological Bulletin 4:* 41 p., 20 pls.

CENOZOIC

FELL, H.B. 1954: Tertiary and Recent Echinoidea of New Zealand: Cidaridae. *N.Z. Geological Survey Paleontological Bulletin 23:* 62 p., 15 pls.

FINLAY, H.J.; MARWICK, J. 1939: The Wangaloan and associated Molluscan faunas of Kaitangata - Green Island Subdivision. *N.Z. Geological Survey Paleontological Bulletin 15:* 140 p., 18 pls.

FLEMING, C.A. 1966: Marwick's illustrations of New Zealand shells, with a checklist of New Zealand Cenozoic Mollusca. *N.Z. Geological Survey Paleontological Bulletin 15: Bulletin 173:* 456 p., 145 pls. (Includes a full bibliography of New Zealand Cenozoic Mollusca.)

GLAESSNER, M.F. 1960: The fossil decapod crustacea of New Zealand and the evolution of the Order Decapoda. *N.Z. Geological Survey Bulletin 31:* 63 p., 7 pls.

HENDERSON, R.A. 1975: Cenozoic spatangoid echinoids from New Zealand. *N.Z. Geological Survey Bulletin 46:* 127 p., 18 pls.

SQUIRES, D.F. 1958: The Cretaceous and Tertiary corals of New Zealand. *N.Z. Geological Survey Paleontological Bulletin 29:* 107 p., 16 pls.

Appendix : Extended Captions

Compiled by I. W. Keyes

These 'extended captions' provide additional information on the specimens figured in the preceding plates which it was not possible to include with the captions. This information, although brief, is intended to provide basic locality and institutional ownership particulars to more fully document the specimens illustrated.

The arrangement of these details follows the same plate and figure sequence given in the preceding pages. After the scientific name of each specimen follows (in the case of the Paleozoic and Mesozoic fossils) an indication of the Class to which the fossil belongs (e.g. bivalve, brachiopod, coral, etc.). For many less experienced users of this handbook, this may not be readily apparent from the fossil illustrated or the name given. In order to illustrate some of the best fossil specimens available many of the illustrations chosen are of type specimens. Their "type-status" (i.e. whether they are primary types - holotypes, lectotypes, neotypes, paratypes, paralectotypes or secondary types - hypotypes) is also given, followed by any relative details concerning the view of the fossil shown. Ownership is indicated by an institutional abbreviation and the following are used.

AU - Auckland University (Geology Department), Auckland.

CM - Canterbury Museum (Geology Department), Christchurch.

NZGS - NZ Geological Survey (Paleontology Section), Lower Hutt.

OM - Otago Museum, Dunedin.

OU - Otago University (Geology Department), Dunedin.

VUW - Victoria University of Wellington (Geology Department).

Following the institutional abbreviation is the individual registration number assigned to the specimen (if allocated) by that institution. Locality details are cited by the inch to the mile map sheet district number (unless otherwise stated) in which the locality occurs, and the fossil record number (of the NZ Fossil File) if it has been assigned. In some instances the official institutional collection registration number also is given (e.g. "GS" or "V") and a brief locality description follows. The age of the specimen is cited (as a New Zealand Stage) followed by the equivalent international geological time scale period.

An abbreviated literature reference follows (see reference list at end) citing any previous figuring and description of the specimen. In the case of type specimens this may refer to the original figuring of the specimen, or may be a reference to a later review work where the specimen was illustrated which would also give references to earlier descriptions and illustration. These references are intended to allow further information on a figured specimen to be obtained by those interested in seeking more details, or further information on the occurrence of the taxon in the New Zealand fossil record.

Plate 1 Representative Cambrian and Ordovician tribolites, brachiopods, and conodonts.

Fig. a. *Nepea* sp. (Trilobite). Cranidium. (NZGS). Loc. S13/f484, Cobb Valley, W Nelson. Late Mid. Cambrian.

Fig. b. *Dorypyge* sp. (Trilobite). Pygidium. (NZGS - AR 476). Loc. S13/f484, Cobb Valley, W Nelson. Late Mid. Cambrian.

Fig. c. *Ptychagnostus* cf. *aculeatus* (Angelin). (Trilobite). Pygidium. (NZGS). Loc. S13/f484, Cobb Valley, W Nelson. Late Mid. Cambrian.

Fig. d. *Koptura* sp. (Trilobite). Three pygidia on surface of slab. (NZGS). Loc. S13/f484, Cobb Valley, W Nelson. Late Mid. Cambrian.

Fig. e. *Grandagnostus* sp. (Trilobite). Pygidium. (NZGS). Loc. S13/f484, Cobb Valley, W Nelson. Late Mid. Cambrian.

Fig. f. *Ptychagnostus* sp. (Trilobite). Cephalon. (NZGS). Loc. S13/f484, Cobb Valley, W Nelson. Late Mid. Cambrian.

Fig. g. *Micromitra* sp. (Inarticulate brachiopod). (NZGS). Loc. S13/f580, Cobb Valley, W Nelson. Late Mid. Cambrian.

Fig. h. *Acrothele* sp. (Inarticulate brachiopod). (NZGS). Loc. S13/f580, Cobb Valley, W Nelson. Late Mid. Cambrian.

Fig. i. *Westergardodina* sp. (Conodont). (NZGS). Loc. S13/f588, Cobb Valley, W Nelson. Late Mid. Cambrian.

Fig. j. *Hirsutodontus hirsutus* Miller. (Conodont). (NZGS). Loc. S19/f684, Crow River, W Nelson. Tremadoc Series (Lr. Ordovician).

Fig. k. *Cordylodus* cf. *angulatus* Pander. (Conodont). (NZGS). Loc. S13/f586, Mount Mytton, W Nelson. Tremadoc Series (Lr Ordovician).

Fig. l. *Eoplacognathus* cf. *reclinatus* (Fahraeus). (Conodont, ambalodiform element). (NZGS). Loc. (metric) M31/f31, Gorge Creek, W Nelson. Llanvirn Series (Mid. Ordovician).

Fig. m. *Periodon aculeatus* Hadding. (Conodont, periodontiform element). (NZGS). Loc. (metric) M31/f31, Gorge Creek, W Nelson. Llanvirn Series (Mid. Ordovician).

Fig. n. *Pygodus* cf. *anserinus* Lamont & Lindstrom. (Conodont, pygodiform element). (NZGS). Loc. (metric) M28/f96, Mount Owen, W Nelson. Upper Llanvirn - Lower Llandeil Series (Mid. Ordovician).

Plate 2 Representative Cambrian trilobites and Lower Ordovician graptolites.

Fig. a. Trilobite limestone with various polymerid trilobites including *Pianaspis* sp. (NZGS - AR 433). Loc. S13/f484; GS4929, Trilobite Rock, Cobb Valley, NW Nelson. (Mid. Cambrian).

Fig. b. *Didymograptus protobifidus* Elles. (NZGS - PR 469). Loc. S2/f490; GS740, Golden Ridge, Aorangi Mine, NW Nelson. *D. protobifidus* zone, Chewtonian (Lr Ordovician).

Fig. c. *Tetragraptus fruticosus* (Hall). Mature rhabdosome. (OU 858, 1 (B399), Locality 13, Benson Collection). Loc. S165/- (unregistered), Cape Providence, SW Otago. Bendigonian (Lr Ordovician).

Fig. d. *Isograptus victoriae maximodivergens* Harris. Hypotype; mature, incomplete rhabdosome. (VUW - VG 46). Loc. S2/f540; V1043, Aorangi Mine, NW Nelson. *I.v. maximodivergens* zone, Yapeenian (Lr Ordovician). [After Cooper, 1979, text-fig. 56].

Fig. e. *Isograptus caduceus imitatus* Harris. Hypotype; immature rhabdosome. (VUW - VG 30). Loc. S2/f552; V1493, Aorangi Mine, NW Nelson. *I.v. maximodivergens* zone, Castlemanian (Lr Ordovician). [After Cooper, 1979, pl. 13f; text-fig. 48b].

Fig. f. *Isograptus victoriae victoriae* Harris. Hypotype; mature rhabdosome. (VUW - VG 43). Loc. S2/f599; V1920, Aorangi Mine, NW Nelson. *I.v. victoriae* zone, Castlemainian (Lr Ordovician). [After Cooper, 1979, text-fig. 53].

Fig. g. *Isograptus victoriae lunatus* Harris. Mature rhabdosome. (VUW - VG 37). Loc. S2/f562; V1564a, Aorangi Mine, NW Nelson. *I.v. lunatus* zone, Castlemainian (Lr Ordovician).

Fig. h. *Tetragraptus approximatus* Nicholson. Hypotype; mature rhabdosome. (VUW - VG 5). Loc. S2/f541; V1044, Aorangi Mine, NW Nelson. *T. approximatus* zone, Lancefieldian (Lr Ordovician). [After Cooper, 1979, pl. 6a; text-fig. 29a].

Fig. i. *Adelograptus? antiquus* (Hall). Hypotype; mature, incomplete, rhabdosome. (CM - zfg 6.2). Loc. S2/f492, Aorangi Mine, NW Nelson. *Adelograptus* zone, Lancefieldian (Lr Ordovician). [After Cooper, 1979, pl. 2d, text-fig. 17a, d].

Plate 3 Representative Lower to Upper Ordovician graptolites and trilobite.

Fig. a. *Glyptograptus* cf. *teretiusculus* (Hisinger). Mature rhabdosome, lengthened by tectonic deformation. (NZGS - PR 470). Loc. S13/f482; GS1231, Cobb Valley, NW Nelson. *G. teretiusculus* zone (= *Diplograptus? decoratus* zone), Upper Darriwilian (Mid. Ordovician).

Fig. b. *Cardiograptus morsus* Harris & Keble. Hypotype; proximal portion of incomplete immature rhabdosome. (NZGS - PR 30). Loc. S2/f556; GS9953, Aorangi Mine, NW Nelson. *Oncograptus* zone, Upper Yapeenian (Lr Ordovician). [After Cooper, 1979, pl. 14f; text-fig. 46a].

Fig. c. *Para ardiograptus* sp. Hypotype; incomplete mature rhabdosome. (NZGS - PR 17). Loc. S2/f545; GS9950, Aorangi Mine, NW Nelson. *Oncograptus* zone, Yapeenian (Lr Ordovician). [After Cooper, 1979, pl. 14a, b; text-fig. 47].

Fig. d. *Paraglossograptus tentaculatus* (Hall). Hypotype; mature rhabdosome. (VUW - VG 72). Loc. S2/f582; V1596, Aorangi Mine, NW Nelson. *P. tentaculatus* zone, Lower Darriwilian (Mid. Ordovician). [After Cooper, 1979, pl. 16e; text-fig. 66g].

Fig. e. *Ogygites collingwoodensis* Reed. (Trilobite). Syntype, specimen lacking librigenae. (NZGS - AR 664). Loc. S3/f527; GS9505, Paturau River, NW Nelson. *Dicellograptus* zone, Gisbornian (Up. Ordovician). [After Reed, 1927, pl. 16, f.1b].

Fig. f. *Oncograptus upsilon biangulatus* Harris & Keble. Hypotype; mature rhabdosome. (NZGS - PR 6). Loc. S2/f540; GS9949, Aorangi Mine, NW Nelson. *Oncograptus* zone, Yapeenian (Lr Ordovician). [After Cooper, 1979, pl. 14j; text-fig. 61a].

Fig. g. *Nemagraptus gracilis* (Hall). Largely complete, mature rhabdosome. (NZGS - PR 471). Loc. S13/f521; GS7373, Cobb Valley, NE Nelson. *N. gracilis* zone (= *Dicellograptus* zone), Gisbornian (Up. Ordovician). [Holotype of *N. gracilis* var. *aotearoaensis* Skwarko, 1962, text-fig. 4, f.15].

Fig. h. *Orthograptus* sp. Mature rhabdosome. (VUW - VG 122). Loc. S19/f620; V1938, Wangapeka Valley, NW Nelson. *Orthograptus* zone, Eastonian (Up. Ordovician).

Plate 4 Representative Lower Devonian fossils.

Fig. a. *Ptychopteria (Actinopteria) mackayi* (Fleming). (Bivalve). Holotype; left valve exterior. (NZGS - TM 2062). Loc. S38/f498; GS130, Lankey's Gully, Reefton, SW Nelson. [After Fleming, 1957a, pl. 14, f.8].

Fig. b. *Fascicostella gervillei* (Defrance). (Brachiopod). Hypotype; natural internal mould of ventral valve. (NZGS - BR 2281). Loc. S19/f491; GS128, Baton River, W Nelson. [After Shirley, 1938, pl. 41, f.6].

Fig. c. *Pleurodictyum megastomum* Dun. (Coral). Hypotype; portion of adult colony. (NZGS - CO 1420). Loc. S13/f491; GS128, Baton River, W Nelson. [After Shirley, 1938, pl. 40, f8].

Fig. d. *Pleurothyrella venusta* Boucot et al. (Brachiopod). Paratype; latex replica of exterior of dorsal valve. (NZGS - BR 1125). Loc. S38/f619; GS6737, Stony Creek, Reefton, SW Nelson. [After Boucot *et al.*, 1963, pl. 27, f.3-5].

Fig. e. *Reeftonella neozelanica* (Allan). (Brachiopod). Holotype; natural internal mould of ventral valve. (NZGS - BR 2282). Loc. S38/f498; GS130, Lankeys Gully, Reefton, SW Nelson. [After Allan, 1935, pl. 2, f.9; 1947, pl. 63,f.1].

Fig. f. *Hipparionyx minor* Clarke. (Brachiopod). Hypotypes; natural internal moulds of two ventral valves. (NZGS - BR 2283, 2284). Loc. S19/f491; GS128, Baton River, W Nelson. [After Shirley, 1938, pl. 42, f.1].

Fig. g. *Tipheophyllum bartrumi* (Allan). (Coral). Hypotypes; transverse section of corallum. (NZGS - CO 1260). Loc. S38/f518; GS5662, Waitahu River, Reefton, SW Nelson. [After Hill, 1956, pl. 1, f.4a].

Fig. h. *Eatonia bithynica* Paeckelmann & Sieverts. (Brachiopod). Hypotype; natural internal mould of ventral valve. (NZGS - BR 2285). Loc. S19/f491; GS128, Baton River, W Nelson. [After Shirley, 1938, pl. 42, f.7].

Fig. i. *Burmeisteria (Digonus)* cf. *expansus* (Hector). (Trilobite). Incomplete thorax and pygidium. (NZGS - AR 676). Loc. S38/f523; GS3737, Rainy Creek, Reefton, SW Nelson.

Fig. j. *Acrospirifer coxi* (Allan). (Brachiopod). Hypotype; natural internal mould of ventral valve. (NZGS - BR 2286). Loc. S38/f523; GS3737, Rainy Creek, Reefton, SW Nelson. [After Grindley *et al.*, 1959, pl. 1, f.6].

Plate 5 Representative Lower and Upper Permian fossils.

Fig. 1. *Attenuatella altilis* Waterhouse. (Brachiopod). Paratypes; latex replica of ventral valve (upper) and dorsal valve. (OU 2446). Loc. S131/f523, E Branch Eglinton River, NW Otago. Mangapirian (Lr Permian). [After Waterhouse, 1968, pl. 2, f.2].

Fig. 2, 3. *Plekonella campbelli* Waterhouse. (Brachiopod). Holotype; latex replica (Fig. 2) and natural internal mould (Fig. 3) of ventral and dorsal valves. (NZGS - BR 438). Loc. S159/f604, GS5885, N end Wairaki Gorge, Southland. Telfordian (Lr Permian). [After Waterhouse, 1964a, pl. 16, f.3, 6, 8].

Fig. 4, 5, 11. *Terrakea brachythaerum* (Morris). (Brachiopod). Hypotypes; Fig. 4, latex replica of dorsal valve. (NZGS - BR 174); Fig. 5, natural internal mould of ventral valve (NZGS - BR 161); Fig. 11, natural internal mould of dorsal valve (NZGS -

BR 163). Loc. S159/f622; GS6071, Letham Burn, Southland. Up. Braxtonian (Up. Permian). [After Waterhouse, 1964a, pl. 14, f.4; pl. 12, f.4; pl. 14, f.11].

Fig. 6. *Horridonia* n.sp. (Brachiopod). Hypotype; leached natural internal mould of ventral valve. (NZGS - BR 51). Loc. S170/f612; GS5078, Waipahi, Southland. Puruhauan (Up. Permian). [After Waterhouse, 1964a, pl. 9, f.1; pl. 8, f.11].

Fig. 7, 8. *Psilocamara saginatum* Waterhouse. (Brachiopod). Paratype; latex replica of ventral and dorsal valves. (NZGS - BR 454). Loc. S159/f604; GS5885, N end Wairaki Gorge, Southland. Telfordian (Lr Permian). [After Waterhouse, 1964a, pl. 19, f.1, 2].

Fig. 9, 12. *Mourlonia impressa* Waterhouse. (Gastropod). Holotype; latex replica. (NZGS - TM 3877). Loc. S15/f513; GS9477, above Croisilles Harbour, E Nelson. Pre Telfordian (Lr Permian). [After Waterhouse, 1966, pl. 1, f.1-3].

Fig. 10, 13. *Aperispirifer wairakiensis* (Waterhouse). (Brachiopod). [Clarke (1979) suggests that *Aperispirifer* could be regarded as a subgenus of *Trigonotreta*.] Paratypes; latex replica of ventral valve and natural internal mould of ventral valve. (NZGS - BR 814, BR 472). Loc. S159/f621; GS6070, Letham Burn, Southland. Lr Braxtonian (Up. Permian). [After Waterhouse, 1964a, pl. 23, f.5; pl. 24, f.3].

Fig. 14, 15. *Walnicholsia subcancellata* (Morris). (Gastropod). Hypotype; natural internal mould. (NZGS - TM 3003). Loc. S159/f622; GS6071, Letham Burn, Southland. Up. Braxtonian (Up. Permian). [After Waterhouse, 1963b, f.29-32].

Fig. 16, 18. *Martinia adentata* Waterhouse. (Brachiopod). Holotype; latex replica of interior of ventral valve and ventral view of natural internal mould. (NZGS - BR 670). Loc. S159/f511; GS5296, opposite junction of Gibraltar Burn and Wairaki River, Southland. Mangapirian (Lr Permian). [After Waterhouse, 1964a, pl. 21, f.10, 11, 13, 14].

Fig. 17. Bryozoan fragments, *Fenestella* sp. (left); *Protoretepora ampla* (Lonsdale), (to right). Hypotypes. (VUW). Loc. S8/-, (unregistered); V1032, Pariwhakaoho Stream, NW Nelson. Up. Braxtonian (Up. Permian). [After Waterhouse & Vella, 1965, pl. 1, f.5].

Plate 6 Representative Upper Permian fossils.

Fig. 1. *Trabeculatia trabeculum* (Waterhouse). (Bivalve). Paratype; natural internal mould. (NZGS - TM 3694). Loc. S15/f512; GS9158, Whangamoa River, E Nelson. Puruhauan. [After Waterhouse, 1963d, pl. 104, f.1; 1979, p. 15].

Fig. 2, 4, 5. *Tomiopsis mantuanensis* (Campbell). Hypotypes; Fig. 2, 4, ventral and dorsal valves (same specimen). Fig. 5, natural internal mould of ventral valve. (NZGS - BR 973, BR 1345). Loc. S159/f478; GS3616, Productus Creek, Southland. Upper Braxtonian. [After Waterhouse, 1968, pl. 11, f.1, 2, 7].

Fig. 3. *Atomodesma obliquatum* Waterhouse. (Bivalve). Holotype; natural internal mould of left valve. (NZGS - TM 3366). Loc. S159/f862; GS7344, first E tributary of Letham Burn, Southland. Lower Braxtonian. [After Waterhouse, 1963d, pl. 100, f.5].

Fig. 6. *Peruvispira imbricata* Waterhouse. (Gastropod). Holotype; latex replica, lateral aspect. (NZGS - TM 3036). Loc. S159/f672; GS6323, first E tributary of Letham Burn, Southland. Upper Braxtonian. [After Waterhouse, 1963c, f.16, 18].

Fig. 7. *Myonia compacta* Waterhouse. (Bivalve). Holotype; right valve. (NZGS - TM 3808). Loc. S159/f478; GS3616, Productus Creek, Southland. Braxtonian. [After Waterhouse, 1969, pl. 15, f.2].

Fig. 8. *Vacunella oblonga* Waterhouse. (Bivalve). Holotype; left valve. (NZGS - TM 3587). Loc. S159/f478; GS3616, Productus Creek, Southland. Braxtonian. [After Waterhouse, 1969, pl. 19, f.6].

Fig. 9. *Myonia* sp. A. Hypotype; left valve. (NZGS - TM 3418). Loc. S159/f862; GS7344, first E tributary of Letham Burn, Southland. Lower Braxtonian. [After Waterhouse, 1969, pl. 13, f.3].

Fig. 10, 11, 12. *Aperispirifer lethamensis* Waterhouse. (Brachiopod). [Clarke (1979) suggests that *Aperispirifer* could be regarded as a subgenus of *Trigonotreta*.] Fig. 10, 12. Holotype; dorsal and ventral valves. (NZGS - BR 1314).

Fig. 11. Hypotype; natural internal mould of immature ventral valve. (NZGS - BR 1313). Loc. S159/f1001; GS9697, second E tributary of Letham Burn, Southland. [After Waterhouse, 1968, pl. 4, f.1-4; pl. 5, f.4].

Plate 7 Representative Upper Permian fossils.

Fig. 1 - 3. *Plekonella iniquitas* Waterhouse. (Brachiopod). Fig. 1. Paratype; natural internal mould of ventral valve. (NZGS - BR 843). Fig. 2. Holotype; natural internal mould of dorsal valve. (NZGS - BR 1236). Loc. S159/f876; GS7358, bluff near head of Productus Creek, Southland. Makarewan. Fig. 3. Hypotype; natural internal mould of dorsal valve, slightly damaged posteriorly. (NZGS - BR 1273). Loc. S159/f886; GS7815, near head of Productus Creek, Southland. Makarewan. [After Waterhouse, 1967, f.11, 13, 15].

Fig. 4, 9. *Durvilleoceras woodmani* Waterhouse. (Ammonoid). Hypotype; juvenile form. Fig. 9. Sutural diagram. (NZGS - CE 2018). Loc. S6/f1; GS7632, Stephens Peninsula, D'Urville Island. Puruhauan. [After Waterhouse 1964b, f.1a, b; 1973, pl. 2, f.b, f.5B].

Fig. 5, 7. *Etheripecten striatura* Waterhouse. (Bivalve). Fig. 5. Paratype; latex replica of right valve. (NZGS - TM 3632). Fig. 7. Holotype; latex replica of left valve. (NZGS - TM 3630). Loc. S170/f612; GS5078, Waipahi, Southland. Puruhauan. [After Waterhouse, 1963e, f.4, 1].

Fig. 6, 10. *Capillonia brevisulcus* (Waterhouse). (Brachiopod). Holotype; leached natural internal mould of ventral valve, and latex cast of exterior of dorsal valve. (NZGS - BR 917). Loc. S171/f498; GS1456, NE of Clinton, Southland. Puruhauan. [After Waterhouse, 1964a, pl. 2, f.4; pl. 3, f.2].

Fig. 8. *Attenuatella incurvata* Waterhouse. (Brachiopod). Holotype; latex replica of dorsal valve. (NZGS - BR 905). Loc. S150/f524; GS7803, Waituna Station, Southland. Puruhauan. [After Waterhouse, 1964a, pl. 20, f.11].

Fig. 11, 13, 14. *Tomiopsis furca* (Waterhouse). (Brachiopod). Holotype; ventral, dorsal and lateral aspects. (NZGS - BR 1194). Loc. S20/f700; GS9140, Roding River, E Nelson. Makarewan. [After Waterhouse, 1967, f. 34, 35, 37].

Fig. 12. *Aperispirifer nelsonensis* (Waterhouse). (Brachiopod). [Clarke (1979) suggests that *Aperispirifer* could be regarded as a subgenus of *Trigonotreta*]. Holotype; ventral valve. (NZGS - BR 500). Loc. S20/f474, GS441, Wairoa Valley, E Nelson. Waiitian. [After Waterhouse, 1964a, pl. 26, f.3].

Fig. 15, 16. *Notospirifer spinosa* Waterhouse. (Brachiopod). Holotype; posterior view with ventral valve uppermost and ventral anterior view. Natural internal moulds. (VUW). Loc. S8/-, (unregistered); V1032, Pariwhakaoho Stream, NW Nelson. Upper Braxtonian. [After Waterhouse & Vella, 1965, pl. 4, f.7, 9].

Fig. 17. *Waagenophyllum novaezelandiae* Leed. (Coral). Holotype; transverse section of part of corallum. (NZGS - CO 1261). Loc. N8/f505, f507; GS5074, Marble Bay, E of Whangaroa Heads, Northland. Puruhauan. [After Leed, 1956, pl. 3, f.5].

Fig. 18. *Wentzelella maoria* Leed. Holotype; transverse section of part of corallum. (NZGS - CO 1264). Loc. N8/f505, f507; GS5074, Marble Bay, E of Whangaroa Heads, Northland. Puruhauan. [After Leed, 1956, pl. 5, f.1].

Plate 8 Endemic New Zealand Permian brachiopods and gastropods.

Fig. 1 - 4. *Coronopsis vagrans* Waterhouse. (Gastropod). Fig. 1, 3, 4; Holotype, latex replica; basal, apical and lateral views. (NZGS - TM 2974). Fig. 2; Paratype, latex replica. (NZGS - TM 2991). Loc. S150/f524; GS7803, SE of Waituna Station, Southland. Upper Braxtonian (Up. Permian). [After Waterhouse, 1963a, f. 26, 31, 33; 27].

Fig. 5 - 7. *Wairakiella rostrata* Waterhouse. (Brachiopod). Paratypes; latex replicas. Fig. 5; ventral valve. (NZGS - BR 1264). Fig. 6, 7; dorsal and lateral views. (NZGS - BR 1272). Loc. S159/f885; GS7814, tributary of Letham Burn, Southland. Makarewan (Up. Permian). [After Waterhouse, 1967, r.16 - 18].

Fig. 8 - 10. *Spinomartinia spinosa* Waterhouse. (Brachiopod). Holotype; natural internal mould: ventral view (Fig. 8), posterior view (dorsal valve uppermost - Fig. 9) and dorsal view (Fig. 10). (OU 2413). Loc. S171/f498, NE of Clinton, Southland. Puruhauan (Up. Permian). [After Waterhouse, 1968, pl. 9, f.2 - 4].

Plate 9 Representative Triassic fossils.

Fig. 1, 2. *Stenopopanoceras coombsi* (Kummel). (Ammonoid). Holotype. (OU 3863). Loc. S179/f674, S of Kaka Point, S Otago. Malakovian (Lr Triassic). [After Kummel, 1965, f.1, 2].

Fig. 3. *Retzia reticulata* Wilckens. (Brachiopod). Paralectotype; latex replica of ventral valve. (NZGS - BR 1072). Loc. S20/f577; GS440, Mt Heslington, E Nelson. Otamitan (Up. Triassic).

Fig. 4. *Retzia morganiana* Wilckens. (Brachiopod). Holotype; natural internal mould of ventral valve. (NZGS - BR 1073). Loc. S20/- (unregistered), Eight-eight Valley, E Nelson. Otamitan (Up. Triassic). [After Wilckens, 1927, pl. 1, f.13a, b].

Fig. 5. *Amphipopanoceras fraseri* (Browne). (Ammonoid). Holotype. (NZGS - CE 172). Loc. S159/f443; GS5199, Beaumont Stream, Southland. Etalian (Mid. Triassic). [After Browne, 1952, pl. 102, f.7, 8; Marwick, 1953, pl. 8, f.1].

Fig. 6. *Owenites* cf. *koeneni* Hyatt & Smith. (Ammonoid). Hypotype. (NZGS - CE 1668). Loc. S159/f574; GS5685, Coal Creek, N Malakoff Hill, Southland. Malakovian (Lr Triassic). [After Kummel, 1959, f.2-4].

Fig. 7. *Amphipopanoceras tepungai* (Browne). (Ammonoid). Holotype. (NZGS - CE 206). Loc. S159/f500; GS3199, Beaumont Stream, Southland. Etalian (Mid. Triassic). [After Browne, 1952, pl. 102, f.14; Marwick, 1953, pl. 8, f.5].

Fig. 8. *Flemingites* cf. *lidacensis* Welter. (Ammonoid). Hypotype. (NZGS - CE 1665). Loc. S159/f574; GS5685, Coal Creek, N Malakoff Hill, Southland. Malakovian (Lr Triassic). [After Kummel, 1959, f.1].

Fig. 9. *Patella nelsonensis* Trechmann. (Gastropod). Holotype; latex replica of shell exterior. (NZGS - TM 2308). Loc. S20/f470; GS436, Eighty-eight Valley, E Nelson. Kaihikuan (Mid. Triassic). [After Trechmann, 1918, pl. 18, f.8a, b; Marwick, 1953, pl. 7, f.3].

Fig. 10, 12, 13. *Mentzeliopsis spinosa* Trechmann. (Brachiopod). Hypotypes. Fig. 10; latex replica of ventral valve (NZGS - BR 1019). Loc. S20/- (unregistered), Wells Creek, E Nelson. Kaihikuan (Mid. Triassic). [After Wilckens, 1927, pl. 3, f.14b]. Fig. 12; latex replica of dorsal valve. (NZGS - BR 1113). Loc. S169/f478; GS381, North Peak, Hokonui Hills, Southland. Kaihikuan (Mid. Triassic). [After Marwick, 1953, pl. 1, f.12]. Fig. 13; natural internal mould of ventral valve. (NZGS - BR 1016). Loc. S170/f632; GS5169, Mountain Road, Waikaka S.D., Southland. Kaihikuan (Mid. Triassic). [After Marwick, 1953, pl. 1, f.13].

Fig. 11, 15, 16, 17. *Daonella apteryx* Marwick. (Bivalve). Fig. 11; Paratype, natural internal mould of left valve. (NZGS - TM 2233). Fig. 15; Holotype, natural internal mould of right valve. (NZGS - TM 2232). [After Marwick, 1953, pl. 6, f.3]. Fig. 16; Paratype, natural internal mould of right valve. (NZGS - TM 2230). Fig. 17; Paratype, natural internal mould of left valve. (NZGS - TM 2231). [After Marwick, 1953, pl. 6, f.2]. Loc. S170/f632; GS5169, Mountain Road, Waikaka S.D., Southland. Kaihikuan (Mid. Triassic).

Fig. 14. *Leiophyllites marshalli* Browne. (Ammonoid). Holotype. (NZGS - CE 158). Loc. S159/f443; GS5199, Beaumont Stream, Southland. Etalian (Mid. Triassic). [After Browne, 1952, pl. 103, f.8; Marwick, 1953, pl. 8, f.12].

Fig. 18, 19. *Spiriferina trechmanni* (Wilckens). (Brachiopod). Holotype; dorsal and ventral views. (NZGS BR 1044). Loc. S72/f498; GS401, Mt Potts, Rangitata Valley, Canterbury. Kaihikuan (Mid. Triassic). [After Trechmann, 1918, pl. 24, f.5; Marwick, 1953, pl. 2, f.9, 14].

Fig. 20, 21. *"Rhynchonella" zealandica* Trechmann. (Brachiopod). Hypotype; natural internal mould ventral and dorsal views. (NZGS - BR 1002). Loc. S160/f483; GS380, Oreti Valley, Southland. Kaihikuan (Mid. Triassic). [After Marwick, 1953, pl. 1, f.4, 5].

Fig. 22. *Dielasma zealandica* Trechmann. (Brachiopod). Hypotype; natural internal mould, dorsal view. (NZGS - BR 1010). Loc. S20/f470; GS436, Eighty-eight Valley, E Nelson. Kaihikuan (Mid. Triassic). [After Marwick, 1953, pl. 1, f.6].

Fig. 23, 24. *Oretia coxi* Marwick. (Bivalve). Fig. 23; Holotype, natural internal mould of right valve. (NZGS - TM 2283). Fig. 24; Paratype, natural internal mould of left valve. (NZGS - TM 2282). Loc. S169/f432; GS378, Benmore Run, Hokonui Hills, Southland. Oretian (Up. Triassic). [After Marwick, 1953, pl. 4, f.2, 4].

Plate 10 Representative Triassic fossils.

Fig. 1, 2. *Proclydonautilus mandevillei* (Marshall). (Ammonoid). Hypotype. (NZGS - CE 134). Loc. S179/f559; GS133, Nugget Point, SE Otago. Otamitan (Up. Triassic). [After Trechmann, 1918, pl. 17, f.3; Marwick, 1953, pl. 9, f.2].

Fig. 3, 4. *Halobia hochstetteri* Mojsisovics. (Bivalve). Hypotypes. Fig. 3; natural internal mould of left valve. (NZGS - TM 5767). Loc. S20/f469; GS435, Eighty-eight Valley, E Nelson. Fig. 4; natural internal mould of right valve. (NZGS - TM 2235). Loc. S179/f473; GS1429, Owaka, SE Otago. Oretian (Up. Triassic). [After Marwick, 1953, pl. 6, f.4, 7].

Fig. 5. *Poroa arata* (Trechmann). (Gastropod). Holotype; latex replica from natural mould of exterior. (NZGS - TM 2315). Loc. S20/f577; GS440, Mt Heslington, E Nelson. Otamitan (Up. Triassic). [After Trechmann, 1918, pl. 18, f.3; Marwick, 1953, pl. 7, f.10].

Fig. 6, 8. *Maoritrigonia nuggetensis* (Trechmann). (Bivalve). Hypotypes. Fig. 6; latex replica of external shell ornament. (NZGS - TM 2789). Loc. S179/f559; GS133, Nugget Point, SE Otago. Fig. 8; latex cast of left valve hinge from natural internal mould. (NZGS - TM 2790). Loc. S179/- (unregistered), "*Trigonia* beds", Nugget Point, SE Otago. Oretian (Up. Triassic). [After Fleming, 1962c, pl. 1, f.5, 7].

Fig. 7, 11. *Hokonuia limaeformis* Trechmann. (Bivalve). Hypotypes. Fig. 7; internal view of left valve. (NZGS - TM 2184). Loc. S20/f497; GS395, between Wairoa Gorge and Eighty-eight Valley, E Nelson. Fig. 11; external view of left valve. (NZGS - TM 2279). Loc. S179/- (unregistered); Nugget Point, SE Otago. Otamitan (Up. Triassic). [After Waterhouse, 1960, pl. 24, f.24 and Marwick, 1953, pl. 3, f.22].

Fig. 9, 13. *Mysidioptera riceae* Waterhouse. (Bivalve). Fig. 9; Paratype, internal view of right valve. (NZGS - TM 2099). Loc. S20/f590; GS442, Wairoa Gorge, E Nelson. Fig. 13; Holotype, latex replica of left valve exterior. (NZGS - TM 2101). Loc. S179/f475; GS1431, Waiwera River, SE Otago. Otamitan (Up. Triassic). [After Waterhouse, 1960b, pl. 23, f.18, 16].

Fig. 10. *Raha spectabilis* (Trechmann). (Gastropod). Holotype; latex replica. (NZGS - TM 2318). Loc. S20/f577; GS440, Mt Heslington, E Nelson. Otamitan (Up. Triassic). [After Trechmann, 1918, pl. 18, f.4; Marwick, 1953, pl. 7, f.12].

Fig. 12. *Cucullaea wellmani* Marwick. (Bivalve). Holotype; right valve interior. (NZGS - TM 2219). Loc. S20/f577; GS440, Mt Heslington, E Nelson. Otamitan (Up. Triassic). [After Marwick, 1953, pl. 3, f.2].

Fig. 14. *Psioidiella conjuncta* (Hector). (Brachiopod). Neotype; natural internal mould. (NZGS - BR 1053). Loc. S160/f501; GS5127, Dipton, Southland. Oretian (Up. Triassic). [After Marwick, 1953, pl. 2, f.13].

Plate 11 Representative Triassic fossils.

Fig. 1, 5. *Clavigera bisulcata* Hector. (Brachiopod). Holotype; dorsal and ventral aspects. (NZGS - BR 1103). Loc. S169/f474; GS371, Benmore Cutting, Oreti Valley, Southland. Otapirian (Up. Triàssic). [After Marwick, 1953, pl. 1, f.23].

Fig. 2, 6. *Clavigera tumida* Hector. (Brachiopod). Holotype; natural internal mould, dorsal and ventral aspects. (NZGS - BR 1108). Loc. S169/f474; GS371, Benmore Cutting, Oreti Valley, Southland. Otapirian (Up. Triassic). [After Marwick, 1953, pl. 1, f.15, 22].

Fig. 3, 4. 7. *Otapiria dissimilis* Cox. (Bivalve). Fig. 3; Hypotype, left valve, exterior. (NZGS - TM 2252). Loc. S169/f464; GS361, Taylors Creek, Hokonui Hills, Southland. Fig. 4 and 7; Neotype; natural internal mould, left and right valves. (NZGS - TM 2254). Loc. S169/f466; GS363, Benmore Yards, Hokonui Hills, Southland. Otapirian (Up. Triassic). [After Marwick, 1953, pl. 3, f.11, 12, 10].

Fig. 8. *Monotis (Entomonotis) richmondiana richmondiana* (Zittel). (Bivalve). Natural internal mould of left valve. (NZGS - TM 5768). Loc. S20/- (unregistered). Brightwater, E Nelson. Warepan (Up. Triassic).

Fig. 9. *Rastelligera elongata* Hector. (Brachiopod). Holotype; latex replica of internal surface of dorsal valve from natural internal mould. (NZGS - BR 1064). Loc. S169/f466; GS363, Benmore Yards, Hokonui Hills, Southland. Otapirian (Up. Triassic). [After Marwick 1953, pl. 2, f.1; Campbell, 1968, pl. 1, f.1a, 1b].

Fig. 10, 13. *Monotis (Entomonotis) richmondiana aucticostata* (Trechmann). (Bivalve). Hypotype; right and left valves. (NZGS - TM 2245). Loc. S20/f550; GS4669, Mt Heslington, E Nelson. Warepan (Up. Triassic). [After Marwick, 1953, pl. 6, f.10].

Fig. 11. *Rastelligera gypaeetus* (Trechmann). (Brachiopod). Holotype; latex replica of dorsal valve. (NZGS - BR 1063). Loc. S169/f471; GS368, South Peak, Benmore, Hokonui Hills, Southland. Otapirian (Up. Triassic). [After Marwick, 1953, pl. 2, f.10; Campbell, 1968, pl. 4,f.1].

Fig. 12, 14. *Mentzelia kawhiana* Trechmann. (Brachiopod). Hypotype; natural internal mould, dorsal and ventral views. (NZGS - BR 1015). Loc. S178/f488; GS470, Mataura River, Southland. Otapirian (Up. Triassic). [After Marwick, 1953, pl. 1, f.10, 11].

Fig. 15. *Rhacophyllites debilis* (Hauer). (Ammonoid). Hypotype. (OU - CCE35). Loc. S159/f677; Taringatura Hills, Southland. Otamitan (Up. Triassic). [After Kummel, 1960, f.5].

Fig. 16. *Kalentera marwicki* Grant-Mackie. (Bivalve). Holotype; latex replica of interior of left valve. (AU - L2608). Loc. N91/f573; Awakino River, N. Taranaki. Otapirian (Up. Triassic). [After Grant-Mackie, 1960, f.4, 6].

Fig. 17, 19. *Pleurotomaria (Sisenna) hectori* Trechmann. (Gastropod). Paratype(?); apical and apertural views. (NZGS - TM 2306). Loc. S169/- (unregistered); Otamita Stream, Hokonui Hills, Southland. Otamitan (Up. Triassic). [After Marwick, 1953, pl. 7, f.1, 2].

Fig. 18. *Monotis (Maorimonotis) calvata* Marwick. (Bivalve). Right valve. (AU - L2513a). Loc. N82/f528; AU1427, Kiritehere, SW Auckland. Warepan (Up. Triassic). [After Grant-Mackie, 1976, pl. 1, f.7].

Fig. 20. *Maoritrigonia* aff. *pirouteti* (Avias). (Bivalve). Hypotype; natural internal mould of right valve. (OU - "JDC 294"). Loc. S159/f772. Dipton Stream, Southland. Otamitan (Up. Triassic). [After Fleming, 1964, f.7"f". (middle)].

Fig. 21. *Maoritrigonia heslingtonensis* (Trechmann). (Bivalve). Lectotype; latex replica of right valve. (NZGS - TM 2221). Loc. S20/- (unregistered); Mt Heslington, E Nelson. Otamitan (Up. Triassic). [After Trechmann, 1918, pl. 22, f.9; Marwick, 1953, pl. 4, f.7].

Fig. 22. *"Minetrigonia"* n.sp. Latex replica of right valve. (NZGS). Loc. S168/f483; GS5219, Taringatura Hills, Southland. Otapirian (Up. Triassic). [After Fleming, 1964, f.7 "a" (top left)].

Plate 12 Representative Southwest Pacific Triassic brachiopods and molluscs.

Fig. 1, 2, 5, 6. *Agonisca corbiensis* Fleming. (Bivalve). Fig. 1, 5; Holotype, latex replica of exterior and interior of right valve from natural moulds. (OU 3814-5). Fig. 2; Paratype, latex replica of left valve. (OU 3813). Fig. 6; Paratype, latex replica of interior of left valve. (OU 3816). Loc. S117/f562, Corbies Creek, N Otago. Kaihikuan (Mid. Triassic). [After Fleming, 1963, f.2, 3, 5, 6].

Fig. 3, 7, 8, 11. *Alipunctifera kaihikuana* (Trechmann). (Brachiopod). Fig. 3; Hypotype, natural internal mould of dorsal valve. (NZGS - BR 1067). Loc. S160/f483; GS380, Oreti Valley, Southland. [After Marwick, 1953, pl. 2, f.15]. Fig. 7; latex replica of ventral valve. (NZGS - BR 2287). Loc. S20/f458; GS389, Wairoa Gorge, E Nelson. Fig. 8, 11. Hypotype, latex replica of interior of dorsal valve, and natural internal mould. (NZGS - BR 1042). Loc. S160/f483; GS380, Oreti Valley, Southland. [After Marwick, 1953, pl. 2, f.17, 19]. Kaihikuan (Mid Triassic).

Fig. 4, 13. *Manticula problematica* (Zittel). (Bivalve). Hypotypes. Fig. 4; latex replica of interior of left valve. (NZGS - TM 2167). Loc. S170/f628; GS5156, S of Waipahi, Southland. [After Waterhouse, 1960b, pl. 21, f.4]. Fig. 13; natural internal mould of right valve. (NZGS - TM 2290). Loc. S20/- (unregistered). Eighty-eight Valley, E Nelson. [After Wilckens, 1927, pl. 2, f.3b; Marwick, 1953, pl. 4, f.1]. Otamitan (Up. Triassic).

Fig. 9, 10, 15. *Mellarium nodulosum* Waterhouse. (Gastropod). Fig. 9, 15; Holotype, latex replica, lateral and apical views. (NZGS - TM 2055). Loc. S159/f754; GS6661, Beaumont Stream, Southland. Fig. 10; Paratype, latex replica, basal view. (NZGS - TM 2056). Loc. S159/f627; GS6075, North Etal Creek, Southland. Etalian (Mid. Triassic). [After Waterhouse, 1960a, f.15-17].

Fig. 12, 16. *Worthenia phillipsi* Waterhouse. (Gastropod). Paratypes, latex replicas lateral and apical views. (NZGS - TM 2786, 2778). Loc. S159/f650; GS6330, Barrett's Hutt, Southland. Etalian (Mid. Triassic). [After Waterhouse, 1961, pl. 1, f.1, 3].

Fig. 14, 17. *Kamupena greggi* Speden. (Gastropod). Holotype; latex replicas, aperture and basal views. (NZGS - TM 2763). Loc. S117/f509; GS7304, Corbies Creek, Otamatapaio River, N Otago. Kaihikuan (Mid. Triassic). [After Gair, Gregg & Speden, 1962, f.19-21].

Fig. 18, 19. *Torastarte bensoni* Marwick. (Bivalve). Hypotypes. Fig. 18; latex replica of interior of left valve. (NZGS- TM 2300). Loc. S169/f470; GS367, Otapiri Creek, Hokonui Hills, Southland. Fig. 19; wax replica of left valve. (NZGS - TM 2299). Loc. S169/- (unregistered). Hokonui Hills, Southland. Otapirian (Up. Triassic). [After Marwick, 1953, pl. 6, f.23, 24].

Fig. 20, 21, 22. *Athyris kaihikuana* Trechmann. (Brachiopod). Hypotypes. Fig. 20; dorsal valve. (NZGS - BR 1094). Loc. S179/f472; GS1427, Kaihiku Range, Southland. Fig. 21, 22; natural internal mould of ventral and dorsal valves. (NZGS - BR 1081). Loc. S160/f483; GS380, Oreti Valley, Southland. Kaihikuan (Mid. Triassic). [After Marwick, 1953, pl. 1, f.14, 19, 20].

Fig. 23, 25 - 27. *Praegonia coombsi* Fleming. (Bivalve). Fig. 23, 25; Holotype, latex replicas of exterior and interior of right valve. (OU - 3719). Loc. S159/f471, White Hill, Taringatura Hills, Southland. Fig. 26, 27; Paratypes, latex replicas of right and left valve interiors. (NZGS - TM 2787). Loc. S169/f636; GS7276, Waimumu Stream, Hokonui Hills, Southland. Kaihikuan (Mid. Triassic). [After Fleming, 1962, pl. 1, f.1-4].

Fig. 24. *Balantioselena gairi* Speden. (Bivalve). Paratype; latex replica of left valve exterior. (NZGS - TM 2776). Loc. S117/f510; GS7204, Corbies Creek, Otamatapaio River, N Otago. Kaihikuan (Mid. Triassic). [After Gair, Gregg & Speden, 1962, f.7].

Fig. 28, 29. *Triaphorus zealandicus* (Trechmann). (Bivalve). Holotype, right and left valves. (NZGS - TM 2298). Loc. S169/- (unregistered), Otamita Stream, Southland. Otamitan (Up. Triassic). [After Marwick, 1953, pl. 3, f.8, 9].

Plate 13 Representative Lower Jurassic fossils.

Fig. 1. *Waehneroceras* n.sp. (Ammonite). Hypotype; latex replica. (NZGS - CE 298). Loc. S169/f455; GS352, Otapiri Creek, Hokonui Hills, Southland. Aratauran (Lr Jurassic). [After Marwick, 1953, pl. 16, f.6].

Fig. 2, 7. *Chlamys (Camptochlamys) wunschae* Marwick. (Bivalve). Latex replicas of left and right valves. Fig. 2; Paratype (NZGS - TM 2407). Loc. S169/f441; GS338, Otapiri Gorge, Hokonui Hills, Southland. Fig. 7; Holotype (NZGS - TM 2406). Loc. S178/f490; GS3203, Wyndham River, E Southland. Ururoan (Lr Jurassic). [After Marwick, 1953, pl. 10, f.23, 24].

Fig. 3. *Pseudolimea inanis* Marwick. (Bivalve). Holotype. (NZGS - TM 2428). Loc. N73/f5; GS1194, N of Ururoa Point, Kawhia, SW Auckland. Ururoan (Lr Jurassic). [After Marwick, 1953, pl. 10, f.10].

Fig. 4, 8. *Otapiria marshalli* (Trechmann). (Bivalve). Hypotypes. Fig. 4; right valve. (NZGS - TM 2390). Loc. S169/f451; GS348, Otapiri Gorge, Hokonui Hills, Southland. Fig. 8; left valve, natural internal mould. (NZGS - TM 2387). Loc. S169/f452; GS349, Otapiri Creek, Hokonui Hills, Southland. Aratauran (Lr Jurassic). [After Marwick, 1953, pl. 11, f7, 8].

Fig. 5. *Dactylioceras* cf. *anginum* (Reinecke). (Ammonite). Hypotype. (NZGS - CE 324). Loc. N73/f5; GS1194, N of Ururoa Point, Kawhia, SW Auckland. Up. Ururoan (Lr Jurassic). [After Marwick, 1953, pl. 16, f.3].

Fig. 6. *Harpoceras* cf. *falcifer* (Sowerby). (Ammonite). Latex replica. (AU - C1012). Loc. N73/f811; S of Albatross Point, Kawhia, SW Auckland. Ururoan (Lr Jurassic).

Fig. 9. *Entolium fossatum* Marwick. (Bivalve). Paratype; latex replica of left valve interior. (NZGS - TM 2401). Loc. S169/f444; GS341, Flag Hill, Hokonui Hills, Southland. Aratauran (Lr Jurassic). [After Marwick, 1953, pl. 13, f.1].

Fig. 10. *Pleuromya urnula* Marwick. (Bivalve). Holotype; left valve. (NZGS - TM 2418). Loc. N73/f5; GS1194, N of Ururoa Point, Kawhia, SW Auckland. Up. Ururoan (Lr Jurassic). [After Marwick, 1953, pl. 10, f.19].

Fig. 11. *Tancredia allani* Marwick. (Bivalve). Holotype; natural internal mould of left valve. (NZGS - TM 2413). Loc. S179/f496; GS3194; Tahatika, Catlins S.D., S Otago. Up. Temaikan (Mid. Jurassic). [After Marwick, 1953, pl. 14, f.4].

Fig. 12, 13. *Pleurotomaria (Perotrochus)* n.sp. (Gastropod). Lateral and apical views. (NZGS - TM 5773). Loc. S184/f604; GS7078, Caberfeidh Quarry, SE Otago. Lr Temaikan (Mid. Jurassic).

Fig. 14, 15. *Brachybelus zieteni* (Werner). (Belemnite). Hypotype; dorsal and ventral view. (AU - C 1049). Loc. N91/f591; AU367, Rauroa Stream, Awakino S.D., SW Auckland. Lr Temaikan (Mid. Jurassic). [After Stevens, 1965, pl. 1, f.1-3].

Fig. 16. *Belemnopsis mackayi* Stevens. (Belemnite). Holotype; latex replica, ventral view. (NZGS - CE 1635). Loc. S169/f692; GS6648, Heale Ridge, Hokonui Hills, Southland. Lr Temaikan (Mid. Jurassic). [After Stevens, 1965, pl. 2, f.1-3].

Fig. 17, 22. *Parainoceramus martini* Speden. (Bivalve). Paratypes. Fig. 17; latex replica of right valve. (AU - L2624a). Loc. N73/f875, Albatross Point, Kawhia, SW Auckland. Fig. 22; natural internal mould of right valve. (NZGS - TM 4065). Loc. N73/f811; GS9674, Albatross Point, Kawhia, SW Auckland. Up. Ururoan (Lr Jurassic). [After Speden, 1970, f.3, 5].

Fig. 18. *Pseudolimea fida* Marwick. (Bivalve). Holotype; natural internal mould of right valve. (NZGS - TM 2429). Loc. S169/f452; GS349, Otapiri Creek, Hokonui Hills, Southland. Aratauran (Lr Jurassic). [After Marwick, 1953, pl. 13, f.15, 16].

Fig. 19. *Trigonia* (s.str.) n.sp. (Bivalve). Hypotype; latex replica of right valve. (NZGS - TM 5776). Loc. S169/f442; GS339, Flag Hill, Hokonui Hills, Southland. Temaikan (Mid. Jurassic). [After Fleming, 1964, f.9"a" (left specimen)].

Fig. 20, 21. *"Rhynchonella" bartrumi* Marwick. (Brachiopod). Holotype; ventral and dorsal valves. (NZGS - BR 1122). Loc. N73/f5; GS1194, N of Ururoa Point, Kawhia, SW Auckland. Up. Ururoan

(Lr Jurassic). [After Marwick, 1953, pl. 15, f.7-10].

Fig. 23, 24. *Inoceramus ururoaensis* Speden. (Bivalve). Fig. 23; Holotype, internal natural mould of right valve. (NZGS - TM 4059). Fig. 24; Paratype, latex replica of right valve. (NZGS - TM 4061). Loc. N73/f815; GS9897, Ururoa Point, Kawhia, SW Auckland. Up. Ururoan (Lr Jurassic). [After Speden, 1970, f.12, 14].

Fig. 25. *Variamussium clamosseum* Marwick. (Bivalve). Holotype; natural internal mould of right valve. (NZGS - TM 2399). Loc. S169/f536; GS5396, Forest Hill S.D., Southland. Lr Temaikan (Mid. Jurassic). [After Marwick, 1953, pl. 10, f.14a, b].

Fig. 26. *Myophorella* n.sp. (Bivalve). Hypotype; latex replica of right valve. (NZGS - TM 5787). Loc. S169/f442; GS339, Flag Hill Hokonui Hills, Southland. Temaikan (Mid. Jurassic). [After Fleming, 1964, f.9, "d"].

Fig. 27. *Meleagrinella* cf. *echinata* (Smith). (Bivalve). Latex replica of specimens crowded on surface of a rock slab. (NZGS). Loc. S184/f495; GS801, Nugget Point coast, SE Otago. Temaikan (Mid. Jurassic).

Fig. 28. *Vaugonia* n.sp. (Bivalve). Hypotype; latex replica of left valve. (NZGS - TM 5786). Loc. S178/f555; GS7120, Rocky Knoll, Dunvegan, E Southland. Temaikan (Mid. Jurassic). [After Fleming, 1964, f.9, "g"].

Fig. 29. *Orthotrigonia* n.sp. (Bivalve). Hypotype; latex replica of left valve. (NZGS - TM 5775). Loc. S178/f528; GS6843, E branch Waipahi River, Southland. Temaikan (Mid. Jurassic). [After Fleming, 1964, f.9, "e"].

Plate 14 Representative Middle and Upper Jurassic fossils.

Fig. 1. *Camptonectes* cf. *laminatus* (Sowerby). (Bivalve). Hypotype; latex replica of left valve. (NZGS - TM 2405). Loc. N73/f18; GS3144, Te Maika, Kawhia, SW Auckland. Temaikan (Mid. Jurassic). [After Marwick, 1953, pl. 10, f.11].

Fig. 2. *Macrocephalites (Kamptokephalites)* cf. *beta-gamma* (Boehm). (Ammonite). (NZGS - CE 2311). Loc. N51/f613; GS11809, Opuatia Stream, Waikato South Head, SW Auckland. Up. Temaikan (Mid. Jurassic).

Fig. 3, 4. *Inoceramus marwicki* Speden. (Bivalve). Fig. 3; Holotype, natural internal mould of right valve. (NZGS - TM 4052). Loc. N91/f636; GS7886, Rauroa Stream, Awakino District, SW Auckland. Fig. 4; Paratype, natural internal mould of left valve. (NZGS - TM 4055). Loc. S184/f507; GS6603, Jack's Bay, Catlins coast, SE Otago. Up. Temaikan (Mid. Jurassic). [After Speden, 1970, f.24, 32].

Fig. 5, 6. *Conodicoelites flemingi* Stevens. (Belemnite). Holotype; ventral and dorsal views. (NZGS - CE 1571). Loc. N73/f550; GS5941, Captain King's, Totara Point, Kawhia, SW Auckland. Heterian (Up. Jurassic). [After Stevens, 1965, pl. 1, f.4-6].

Fig. 7, 8. *Haastina haastiana* (Wilckens). (Bivalve). Fig. 7; slab with internal natural moulds. (NZGS). Fig. 8; Holotype, latex replica of right valve interior. (NZGS - TM 2415). Loc. S184/- (unregistered), prob. Jack's Bay, Catlins coast, SW Otago. Up. Temaikan (Mid. Jurassic). [After Wilckens, 1927, pl. 8, f.10a, b; Marwick 1953, pl. 3, f.3, 4].

Fig. 9. *Vaugonia kawhiana* (Trechmann). (Bivalve). Latex replica of right valve. (AU - L 2467). Loc. N91/f595; AU371. Palmer Creek, Awakino Valley, SW Auckland. Heterian (Up. Jurassic).

Fig. 10, 11. *Pleuromya milleformis* Marwick. (Bivalve). Holotype and Paratype; natural internal moulds of right valves. (NZGS - TM 2431, 2419). Loc. N73/f18; GS3144, Te Maika, Kawhia, SW Auckland. Temaikan (Mid. Jurassic). [After Marwick, 1953, pl. 13, f.6, 7].

Fig. 12. *Kossmatia* cf. *desmidoptycha* Uhlig. (Ammonite). (NZGS - CE 1246). Loc. N73/f606; GS6199, Lemon Point, Kawhia, SW Auckland. Heterian (Up. Jurassic).

Fig. 13, 15, 21. *Malayomaorica malayomaorica* (Krumbeck). (Bivalve). Fig. 13; right valve. (NZGS - TM 5789). Loc. N73/f563; GS5954, Heteri Promontory, Kawhia, SW Auckland. Heterian (Up. Jurassic). Fig. 15; Hypotype, right valve. (NZGS - TM 3937). Loc. N73/f558; GS5949, Heteri Promontory, Kawhia, SW Auckland.

Heterian (Up. Jurassic). [After Marwick, 1966, f.13]. Fig. 21; right valve interior. (NZGS - TM 5783). Loc. N73/f548; GS5914, Heteri Promontory, Kawhia, SW Auckland. Heterian (Up. Jurassic).

Fig. 14. *Paraboliceras* aff. *sabineanum* (Oppel). (Ammonite). (NZGS - CE 1403). Loc. N73/f597; GS6193, Mairi Point, Kawhia, SW Auckland. Ohauan (Up. Jurassic).

Fig. 16, 17. *Kutchithyris hendersoni* Marwick. (Brachiopod). Holotype; ventral and dorsal views. (NZGS - BR 1119). Loc. N73/f27; GS3143, Totara Peninsula, Kawhia, SW Auckland. Heterian (Up. Jurassic). [After Marwick, 1953, pl. 15, f.13, 14].

Fig. 18. *Pinna kawhiana* Marwick. (Bivalve). Holotype. (NZGS - TM 2404). Loc. N73/f27; GS3143, Totara Peninsula, Kawhia, SW Auckland. Heterian (Up. Jurassic). [After Marwick, 1953, pl. 13, f.14].

Fig. 19, 20. *Cerithinella* sp. (Gastropod). Fig. 19; Hypotype. (NZGS - TM 2439). Loc. N73/f27; GS520, Totara Peninsula, Kawhia, SW Auckland. [After Marwick, 1953, pl. 13, f.2]. Fig. 20; (NZGS - TM 5784). Loc. N73/f547; GS5913, Heteri Promontory, Kawhia, SW Auckland. Heterian (Up. Jurassic).

Fig. 22. *Inoceramus brownei* Marwick. (Bivalve). Holotype; natural internal mould. (NZGS - TM 2372). Loc. S183/f492; GS2998, Quarry Hills, Waikawa S.E., Southland. Lr Temaikan (Mid. Jurassic). [After Marwick, 1953, pl. 13, f.17].

Fig. 23. *Grammatodon (Indogrammatodon) taylori* Marwick. (Bivalve). Holotype; left valve. (NZGS - TM 2367). Loc. N73/f30; GS1193, Heteri Promonotry, Kawhia, SW Auckland. Lr Temaikan (Mid. Jurassic). [After Marwick, 1953, pl. 11, f.3].

Fig. 24, 26. *Cryptorhynchia kawhiana* (Trechmann). (Brachiopod). Hypotype; ventral and dorsal aspects. (NZGS - BR 1118). Loc. N73/f27; GS1190, Totara Peninsula, Kawhia, SW Auckland. Heterian (Up. Jurassic). [After Marwick, 1953, pl. 15, f.12].

Fig. 25. *Astarte morgani* (Trechmann). (Bivalve). Hypotype; right valve. (NZGS - TM 2422). Loc. N73/f27; GS1190, Totara Peninsula, Kawhia, SW Auckland. Heterian (Up. Jurassic). [After Marwick, 1953, pl. 10, f.22, 25].

Fig. 27. *Hibolithes catlinensis* (Hector). (Belemnite). Lectotype; ventral view. (NZGS - CE 568). Loc. S184/f491; GS21, Cannibal Bay, SE Otago. Up. Temaikan (Mid. Jurassic). [After Stevens, 1965, pl. 14, f.1, 2].

Fig. 28. *Myophorella* n.sp. (Bivalve). Hypotype; left valve. (NZGS - TM 5785). Loc. N73/f546; GS5912, W of Heteri Promontory, Kawhia, SW Auckland. Heterian (Up. Jurassic). [After Fleming, 1964, f.9, "c"].

Fig. 29. *Belemnopsis aucklandica trechmanni* Stevens. (Belemnite). Holotype; ventral view. (NZGS - CE 1209). Loc. N73/f597; GS6193, Mairi Point, Kawhia, SW Auckland. Ohauan (Up. Jurassic). [After Stevens, 1965, pl. 9, f.1-3].

Fig. 30. *Belemnopsis keari* Stevens. (Belemnite). Holotype; ventral view. (NZGS - CE 1722). Loc. N82/f87; GS2525, quarry Pomarangei Road, Marakopa, SW Auckland. Heterian (Up. Jurassic). [After Stevens, 1965, pl. 2, f.6-8].

Plate 15 Representative Middle and Upper Jurassic fossils.

Fig. 1. *Taeniopteris crassinervis* (Feistmantel). Hypotype; a complete frond. (NZGS). Loc. S178/f496; GS406, Mataura Falls, Southland. Up. Temaikan (Mid. Jurassic). [After Arber, 1917, pl. 10, f.1].

Fig. 2 - 5. *Buchia* aff. *subpallasi* (Krumbeck). (Bivalve). Hypotypes. Fig. 2; left valve. (NZGS - TM 5777). Loc. S62/f529, GS502, W slope Cheviot Hills, N Canterbury. Fig. 3, 4, 5; left valve, lateral, dorsal and posterior aspects (same specimen). (NZGS - TM 5778). Loc. S41/f531; GS5828, Gore River, Clarence Valley, Marlborough. ?Puaroan (Up. Jurassic). [After Fleming, 1958, f.2, 3-5].

Fig. 6, 7. *Buchia* aff. *misolica* (Krumbeck). (Bivalve). Fig. 6; latex replica of left valve external. (NZGS). Loc. N73/- (unregistered), Puti Point, Kawhia, SW Auckland. Puaroan (Up.

Jurassic). Fig. 7; right valve. (NZGS - TM 5782). Loc. N73/f631; GS6480, Puti Point, Kawhia, SW Auckland. Puaroan (Up. Jurassic).

Fig. 8 - 12. *Buchia plicata* (Zittel). (Bivalve). Hypotypes. Fig. 8, 9; lateral and posterior dorsal views of left valve. (AU - L 2603). Loc. N51/f636; Waikato South Head, SW Auckland. Fig. 10 - 12; posterior dorsal, and right and left valve views. (AU - L 2604). Loc. N51/f650; Huriwai River, Waikato South Head, SW Auckland. Puaroan (Up. Jurassic). [After Fleming, 1959, f.3a, b; 4a-c].

Fig. 13 - 15, 19, 20. *Buchia hochstetteri* Fleming. (Bivalve). Fig. 13; Paratype, right valve. (NZGS - TM 2176). Loc. N51/f12; GS5068, Kaawa Creek Road, Port Waikato, SW Auckland. Fig. 14, 19, 20; Holotype, posterior dorsal, right and left valve views. (NZGS - TM 2174). Loc. N51/- (unregistered), Waikato Heads, SW Auckland. Fig. 15; Paratype, right valve. (NZGS - TM 2175). Loc. N51/- (unregistered), Maretae Valley, Port Waikato, SW Auckland. Puaroan (Up. Jurassic). [After Fleming, 1959, f.19a, b; 17a-c; 18a-c].

Fig. 16. *Hibolithes arkelli* Stevens. (Belemnite). Holotype; ventral view. (NZGS - CE 1759). Loc. N73/f628; GS6475, Puti Point, Kawhia, SW Auckland. Puaroan (Up. Jurassic). [After Stevens, 1965, pl. 17, f.1].

Fig. 17. *Hibolithes marwicki marwicki* Stevens. (Belemnite). Holotype; vental view. (NZGS - CE 1788). Loc. N73/f631; GS6480, Point West, Kawhia, SW Auckland. Puaroan (Up. Jurassic). [After Stevens, 1965, pl. 19, f.1].

Fig. 18. *Aulacosphinctoides brownei* (Marshall). (Ammonite). (AU - C 376). Loc. N51/f643, Maretae Valley, Port Waikato, SW Auckland. Puaroan (Up. Jurassic).

Plate 16 Representative Upper Jurassic fossils.

Fig. 1. *Sphenopteris* sp. (NZGS). Loc. N51/- (unregistered), Port Waikato, SW Auckland. Puaroan (Up. Jurassic).

Fig. 2. *Anopaea* n.sp. (Bivalve). Hypotype; right valve. (OU - 3810). Loc. S62/f526; E of Mt Beautiful, Cheviot Hills, N Canterbury. Puaroan. [After Fleming, 1958, f.14].

Fig. 3. *Microphyllopteris pectinata* Arber ex Hector. (NZGS). Loc. N51/- (unregistered), Port Waikato, SW Auckland. Puaroan (Up. Jurassic).

Fig. 4. *Grammatodon (Indogrammatodon) fyfei* Marwick. (Bivalve). Holotype; left valve. (NZGS - TM 2365). Loc. N51/f516; GS5115, Waikato South Head, SW Auckland. Puaroan (Up. Jurassic). [After Marwick, 1953, pl. 11, f.1].

Fig. 5. *Uhligites motutaranus* (Boehm). (Ammonite). (NZGS - CE 1444). Loc. N73/f625; GS6474, Motutara Bluff, Kawhia, SW Auckland. Puaroan (Up. Jurassic).

Fig. 6. *Psilotrigonia* n.sp. (Bivalve). Right valve. (OU - CP41). Loc. S62/f534, Jed River, N Canterbury. Puaroan (Up. Jurassic).

Fig. 7. *Myophorella* n.sp. (Bivalve). Hypotype; right valve. (NZGS - TM 5788). Loc. N51/f650; GS6362, Huriwai River, Waikato Heads. Puaroan (Up. Jurassic). [After Fleming, 1964, f.9, "b"].

Fig. 8, 9, 11. *Otapiria masoni* Marwick,. (Bivalve). Fig. 8; internal natural mould of left valve. (NZGS). Loc. N51/- (unregistered), Port Waikato, SW Auckland. Puaroan (Up. Jurassic). Fig. 9; Paratype, natural internal mould of right valve. (NZGS - TM 2388). Loc. N51/- (unregistered), Maretai Stream, Waikato Heads, SW Auckland. Puaroan (Up. Jurassic). [After Marwick, 1953, pl. 11, f.11]. Fig. 11; Holotype, natural internal mould of left valve. (NZGS - TM 2389). Loc. N51/- (unregistered), Maretai Stream, Waikato Heads, SW Auckland. Puaroan (Up. Jurassic). [After Marwick, 1953, pl. 11, f.10].

Fig. 10. *Nucula (Palaeonucula) putiensis* Marwick. (Bivalve). Holotype; right valve exterior. (NZGS - TM 2360). Loc. N73/f26; GS1019, Puti Point, Kawhia, SW Auckland. Puaroan (Up. Jurassic). [After Marwick, 1953, pl. 10, f.4].

Fig. 12. *Torquatisphinctes* n.sp. (Ammonite). (NZGS - CE 1455). Loc. N73/f14; GS522, Motutara Bluff, Kawhia, SW Auckland. Puaroan (Up. Jurassic).

Fig. 13. *Inoceramus* aff. *everesti* Oppel. (Ammonite). Hypotype; natural internal mould of right valve. (NZGS - TM 2371). Loc. N51/f12; GS5068, Kaawa Creek Road, Port Waikato, SW Auckland. Puaroan (Up. Jurassic). [After Marwick, 1953, pl. 12, f.4].

Fig. 14, 15. *Belemnopsis aucklandica aucklandica* (Hochstetter). (Belemnite). Hypotype; ventral and dorsal views. (NZGS - CE 410). Loc. N51/f12; GS5068, Kaawa Creek Road, Port Waikato, SW Auckland. Puaroan (Up. Jurassic). [After Stevens, 1965, pl. 8, f.1-3].

Plate 17 Representative Southwest Pacific Jurassic brachiopods and molluscs.

Fig. 1, 2. *Pseudaucella marshalli* (Trechmann). (Bivalve). Hypotypes. Fig. 1; left valve. (NZGS - TM 2394). Loc. N73/f34; GS1191, Te Maika Peninsula, Kawhia, SW Auckland. Fig. 2; (NZGS - TM 2391). Loc. N91/f9; GS915, Lindross Creek, Awakino Gorge, SW Auckland. Ururoan (Lr Jurassic). [After Marwick, 1953, pl. 10, f.17; pl. 14, f.3].

Fig. 3, 6. *Clavigera cuneiformis* Hector. (Brachiopod). Hypotype; dorsal and ventral valves. (NZGS - BR 1116). Loc. S169/f461; GS358, Taylor's Creek, Hokonui Hills, Southland. Aratauran (Lr Jurassic). [After Trechmann, 1918, pl. 25, f.4a, b; Marwick, 1953, pl. 15, f.11].

Fig. 4, 7, 9 - 11. *Kalentera mackayi* Marwick. (Bivalve). Fig. 4, 7, 9, 11. Holotype; natural internal mould of both valves, with latex replica of the left valve interior. (NZGS - TM 2420). Loc. S169/f452; GS349, Otapiri Creek, Hokonui Hills, Southland. Fig. 10; Paratype, natural internal mould; latex replica of right valve interior. (NZGS - TM 2416). Loc. S169/f447; GS344, Flag Hill, Hokonui Hills, Southland. Aratauran (Lr Jurassic). [After Marwick, 1953, pl. 13, f.5, 8, 9, 11, 12].

Fig. 5, 8. *Kalentera flemingi* Marwick. (Bivalve). Paratypes; left valve, and natural internal mould of left valve. (NZGS - TM 5769, 5770). Loc. N91/f8; GS983, Awakino Valley Road, SW Auckland. Up. Ururoan (Lr Jurassic).

Fig. 12, 14, 15. *Sphaeriola leedae* Marwick. (Bivalve). Fig. 12, 15. Paratype; latex replica of right valve, and internal natural mould of right valve (same specimen). (NZGS - TM 2412, 2425). Fig. 14; Holotype, internal natural mould of left valve. (NZGS - TM 2410). Loc. S169/f452, GS349, Otapiri Creek, Hokonui Hills, Southland. Aratauran (Lr Jurassic). [After Marwick, 1953, pl. 10, f.12, 16].

Fig. 13. *Spiriferina radiata* Hector. (Brachiopod). Neotype; natural internal mould of ventral valve. (NZGS - BR 1121). Loc. S169/f440; GS337, Otapiri Gorge, Hokonui Hills, Southland. Ururoan (Lr Jurassic). [After Marwick, 1953, pl. 15, f.19, 20].

Fig. 16. *Inoceramus* cf. *subhaasti* Wandel. (Bivalve). Natural internal mould of right valve. (NZGS - TM 5774). Loc. N73/f12; GS9937, Old Kihi Road, Hauturu, Kawhia, SW Auckland. Heterian (Up. Jurassic).

Fig. 17. *Inoceramus galoi* Boehm. (Bivalve). Hypotype; natural internal mould of right valve. (NZGS - TM 4070). Loc. N91/f64; GS1594, junction Mangaotaki and Mokau Rivers, SW Auckland. Heterian (Up. Jurassic). [After Speden, 1970, f.39].

Fig. 18, 20. *Inoceramus haasti* Zittel. (Bivalve). Natural internal moulds of left valves. Fig. 18; (NZGS - TM 5771). Loc. N73/f11; GS521, Kowhai Point, Kawhia, SW Auckland. Fig. 20; (NZGS - TM 5772). Loc. N74/- (unregistered), quarry Kinohaku Saddle, SW Auckland. Ohauan (Up. Jurassic).

Fig. 19. *Spiriferina ongleyi* Marwick. (Brachiopod). Holotype; natural internal mould of ventral valve. (NZGS - BR 1117). Loc. S169/f446; GS343, Flag Hill, Hokonui Hills, Southland. Aratauran (Lr Jurassic). [After Marwick, 1953, pl. 15, f.18].

Plate 18 Some fossils from the Torlesse Supergroup (Triassic and Jurassic).

Fig. 1. *Titahia corrugata* Webby. (Annelid - "Tube worm"). Holotype (main upper tube) and Paratypes. (NZGS). Loc. N160/f538; GS6973, Rock Point, SW Wellington. Late Triassic. [After Webby, 1958, f.1].

Fig. 2, 3. *Halorella ruahinensis* Campbell. (Brachiopod). Holotype; dorsal and lateral views. (VUW - VB 10). Loc. N140/f519, Umutoi Stream, Oroua Valley, Ruahine Range, NE Wellington. Otamitan - Warepan (Up. Triassic). [After Milne & Campbell, 1969, f.1a-d].

Fig. 4. *Torlessia (= Terebellina) mackayi* (Bather). (Annelid - "tube worm"). Hypotype; slab with several flattened specimens. (NZGS). Loc. N160/f621; GS9741, Owhariu Bay, SW Wellington. ?Kaihikuan - Otamitan (Mid. to Up. Triassic). [After Fleming, 1970, pl. 9, f.c].

Fig. 5, 9. *Holcothyris (?) kaiwaraensis* Campbell. (Brachiopod). Holotype; anterior and dorsal views. (OU - C 1710). Loc. S62/f518, Kaiwarra Valley, N Canterbury. Late Jurassic. [After Campbell, 1965, pl. 1, f.7, 8].

Fig. 6, 7, 8. *Burmirhynchia warreni* Campbell. (Brachiopod). Holotype; dorsal, ventral and anterior views. (OU - C 1700). Loc. S62/f506, Kaiwara Valley, N Canterbury. Late Jurassic. [After Campbell, 1965, pl. 1, f.1-5].

Plate 19 Representative Lower Cretaceous fossils.

Fig. 1. *Inoceramus warakius* Wellman. (Bivalve). Holotype; section of adult shell (external). (NZGS - TM 2118). Loc. N71/f574; GS6660, Wairongomai Stream, Tapuwaeroa Valley, N Gisborne. ?Urutawan - Motuan. [After Wellman, 1959, pl. 12, f.9].

Fig. 2 - 4. *Aucellina* cf. *radiatostriata* Bonarelli. (Bivalve). Fig. 2, 4; natural internal mould of left valve. (NZGS - TM 4964). Loc. N70/f501; GS5773, coast N of Hawai River, E Bay of Plenty. Korangan. Fig. 3; latex replica of left valve exterior. (NZGS - TM 4965). Loc. N88/f493; GS1403, head of Elmar and Harding Streams, Koranga, Gisborne District. Korangan.

Fig. 5. *Trigonia* n.sp. cf. *tenuis* Kitchin. (Bivalve). Left valve. (NZGS - TM 4966). Loc. N88/f495; GS3154, NW of Williams Road, Koranga, Gisborne District. Korangan.

Fig. 6. *Maccoyella* cf. *reflecta* (Moore). [= *M. magnata* Marwick]. (Bivalve). Holotype; left valve. (NZGS - TM 2121). Loc. N88/f494; GS1989, track N of Koranga River, Gisborne District. Korangan. [After Marwick, 1939, pl. 61, f.1, 2; Day & Speden, 1970, f.4].

Fig. 7 - 10. *Aucellina* cf. *gryphaeoides* (Sowerby). (Bivalve). Fig. 7, 9; internal natural moulds of left valve, and a right valve. (NZGS - TM 4967, 4969). Loc. N87/f555; GS9757, junction of Winding and Kerikeri Streams, W Gisborne. Urutawan. Fig. 8, 10; internal natural mould of left valve, and latex replica of right valve exterior. (NZGS - TM 4968, 4970). Loc. N88/f519; GS5792, Koranga Stream, Gisborne District. Urutawan.

Fig. 11. *Inoceramus kapuus* Wellman. (Bivalve). Holotype; juvenile shell, exterior. (NZGS - TM 2120). Loc. N88/f512; GS5786, Motu Section, Gisborne District. Urutawan. [After Wellman, 1959, pl. 12, f.11].

Fig. 12 - 14. *Aucellina euglypha* Woods. (Bivalve). Fig. 12, 13; right valve. (NZGS - TM 4971). Loc. S35/f606; GS6118, Bride Creek, Clarence River Headwaters, Marlborough. Motuan. Fig. 14; latex replica of left valve exterior. (NZGS - TM 4972). Loc. N88/f675; GS8404, Ron Stream, Motu River, Gisborne District. Motuan.

Fig. 15. *Inoceramus ipuanus* Wellman. (Bivalve). Holotype; latex replica of right valve. (NZGS - TM 2115). Loc. S48/f516; GS5830, trib. of Seymour Stream, Clarence Valley, Marlborough. Motuan. [After Wellman, 1959, pl. 11, f.6].

Fig. 16, 17. *Wellmanites zelandicus* Wright. (Ammonite). Holotype; whorl frgment with shell removed, lateral and peripheral views. (NZGS - CE 1065). Loc. N158/f541; GS6090, Gentle Annie Stream, E Wellington. Motuan. [After Wright, 1957, pl. 54, f.3a-c].

Fig. 18, 19. *Dimitobelus superstes* (Hector). (Belemnite). Hypotype; lateral and ventral views. (NZGS - CE 533). Loc. S35/f106; GS741, Upper Awatere Valley, Marlborough. Motuan. [After Stevens, 1965, pl. 22, f.4-6].

Plate 20 Representative Lower and Upper Cretaceous fossils.

Fig. 1. *Otoscaphites awanuiensis* Wright. (Ammonite). Holotype. (NZGS - CE 1379). Loc. N72/f506; GS6511, Waiotautu Stream, Port Awanui, N Gisborne. Ngaterian. [After Wright, 1957, pl. 54, f7a, b; Henderson, 1973, Fig. 10.9].

Fig. 2. *Eselaevitrigonia meridiana* (Woods). (Bivalve). Lectotype; right valve. (NZGS - TM 2469). Loc. S35/f106; GS741, upper Awatere Valley, Marlborough. Ngaterian. [After Woods, 1917, pl. 1, f.3a, b].

Fig. 3, 4. *Sciponoceras baculoide* (Mantell). (Ammonite). Hypotype; ventral and lateral views. (VUW - C 378). Loc. S35/f881, Sawpit Gully, Coverham, Marlborough. Ngaterian. [After Henderson, 1973, Fig. 6.5a-c].

Fig. 5. *Inoceramus hakarius* Wellman. (Bivalve). Holotype; right valve. (NZGS - TM 2117). Loc. N71/f553; GS6521, Mata River, near Waingakia Stream mouth, N Gisborne. Ngaterian. [After Wellman, 1959, pl. 11,f.8].

Fig. 6. *Inoceramus fyfei* Wellman. (Bivalve). Holotype; natural internal mould of right valve. (NZGS - TM 2114). Loc. N88/f539; GS6277, Mill Road, Motu Falls Section, N Gisborne. Ngaterian. [After Wellman, 1959, pl. 11, f.5].

Fig. 7. *Scaphites equalis coverhamensis* Henderson. (Ammonite). Paratype. (VUW - C 389). Loc. S35/f881, Sawpit Gully, Coverham, Marlborough. Ngaterian. [After Henderson, 1973, Fig. 10.7].

Fig. 8 - 10. *Inoceramus tawhanus* Wellman. (Bivalve). Fig. 8, 10; right valve exterior and lateral view. (NZGS - TM 4973). Loc. S35/f630; GS6534, Limestone Creek, Awatere Valley, Marlborough. Ngaterian. Fig. 9; Hypotype, natural internal mould of left valve. (NZGS - TM 2740). Loc. S35/f106; GS741, Upper Awatere Valley, Marlborough. Ngaterian. [After Woods, 1917, pl. 4, f.3].

Fig. 11, 12. *Inoceramus concentricus* Parkinson. (Bivalve). Hypotype; left valve exterior and lateral view. (NZGS - TM 2743). Loc. S35/- (unregistered), Cover Creek, Coverham, Marlborough. Ngaterian. [After woods, 1917, pl. 3, f.10a, b].

Plate 21 Representative Upper Cretaceous fossils.

Fig. 1, 2. *Inoceramus rangatira* Wellman. (Bivalve). Fig. 1, Holotype; section of adult external shell. (NZGS - TM 2113). Loc. N79/f503; GS5993, Mangaotane River, N Gisborne. Arowhanan. [After Wellman, 1959, pl. 10, f.4]. Fig. 2; Topotype, section of adult external shell. (NZGS - TM 4974). Loc. N79/f503; GS5993, Mangaotane River, N Gisborne. Arowhanan.

Fig. 3. *Iotrigonia glyptica* (Woods). (Bivalve). Lectotype; right valve. (NZGS - TM 2480). Loc. S41/f494; GS570, Seymour River, Clarence Valley, S Marlborough. Arowhanan. [After Woods, 1917, pl. 1, f.8a, b].

Fig. 4, 5. *Inoceramus bicorrugatus* Marwick. (Bivalve). Fig. 4, Holotype; section of adult external shell. (NZGS - TM 4975). Loc. N79/f8; GS10893, Mangaotane River, N Gisborne. Mangaotanean. [After Marwick, 1926b, f.1; Grindley et al., 1959, pl. 3, f.4]. Fig. 5, Hypotype; section of external shell showing change of sculpture. (NZGS - TM 4976). Loc. N79/f504; GS5992, Mangaotane River, N Gisborne. Mangaotanean. [After Grindley *et al.*, 1959, pl. 3, f.5].

Fig. 6, 7. *Inoceramus nukeus* Wellman. (Bivalve). Fig. 6; right valve. (NZGS - TM 4977). Loc. S35/f98; GS518, Coverham, Clarence Valley, Marlborough. Teratan. Fig. 7; Holotype, natural internal mould of left valve. (NZGS - TM 2111). Loc. N155/f498, GS4451, Waimata River, Cape Turnagain, SW Hawkes Bay. Teratan. [After Wellman, 1959, pl. 10,f.2]

Fig. 8, 9. *Dimitobelus? ongleyi* Stevens. (Belemnite). Holotype; left lateral and dorsal views. (NZGS - CE 1900). Loc. N159/f449; GS1816, trib. of Marauhika River, Castlepoint S.D., E Wellington. ?Teratan. [After Stevens, 1965, pl. 24, f.4-7].

Fig. 10. *Inoceramus opetius* Wellman. (Bivalve). Holotype; right valve exterior. (NZGS - TM 2112). Loc. N142/f504, GS3225, Waimarama, S Hawkes Bay. Teratan. [After Wellman, 1959, pl. 10, f.3].

Plate 22 Representative Upper Cretaceous fossils.

Fig. 1. *Inoceramus pacificus* Woods. (Bivalve). Lectotype; right valve. (NZGS - TM 2739). Loc. S55/f25; GS13, Haumuri Bluff, S Marlborough. Piripauan. [After Woods, 1917, pl. 14, f.2].

Fig. 2. *Inoceramus australis* Woods. (Bivalve). Lectotype; right valve. (NZGS - TM 2738). Loc. S55/f25; GS13, Haumuri Bluff, S Marlborough. Piripauan. [After Woods, 1917, pl. 3, f.3].

Fig. 3. *Pterotrigonia pseudocaudata* Hector. (Bivalve). Neotype; right valve. (NZGS - TM 2474). Loc. S55/f25; GS13, Haumuri Bluff, S Marlborough. Piripauan. [After Woods, 1917, pl. 8, f.2].

Fig. 4. *Kossmaticeras (K.) haumuriensis* (Hector). (Ammonite). Lectotype, (NZGS - CE 890). Loc. S56/f25; GS2, Haumuri Bluff, S Marlborough. Piripauan. [After Woods, 1917, pl. 19, f.5a-c; Henderson, 1970, pl. 4, f.3a, b].

Fig. 5, 6. *Dimitobelus lindsayi* (Hector). (Belemnite). Lectotype left lateral and ventral view. (NZGS - CE 1078). Loc. S55/f25; GS13, Haumuri Bluff, S Marlborough. Piripauan. [After Woods, 1917, pl. 20, f.7a, b; Stevens, 1965, pl. 25, f.7-9].

Fig. 7. *Inoceramus matotorus* Wellman. (Bivalve). Holotype; anterior view of both valves. (NZGS - TM 2110). Loc. N80/f489; GS1604, Ihungia Stream, N Gisborne. Haumurian. [After Wellman, 1959, pl. 10, f.1].

Fig. 8. *Ostrea lapillicola* Marwick. (Bivalve). Holotype, right valve exterior. (NZGS - TM 2506). Loc. N72/- (unregistered), Tuparoa Beach, N Gisborne. Haumurian. [After Marwick, 1926b, f.4-6].

Fig. 9. *Perissoptera waiparaensis* (Hector). (Gastropod). Hypotype. (NZGS - TM 2628). Loc. S56/f31; GS9, Haumuri Bluff, S Marlborough. Haumurian. [After Wilckens, 1922, pl. 2, f.9].

Fig. 10, 11. *Dimitobelus hectori* Stevens. (Belemnite). Fig. 10, Hypotype, dorsal (or ventral) view. (NZGS - CE 1086). Loc. S163/f490; GS5152, Brighton, Otago. Haumurian. Fig. 11, Holotype, lateral view. (NZGS - CE 1141). Loc. N15/f491, GS728, Waitangi River, Bay of Islands, Northland. Haumurian. [After Stevens, 1965, pl. 25, f.17; 21].

Plate 23 Representative Upper Cretaceous fossils.

Fig. 1. *Conchothyra parasitica* (Hutton). (Gastropod). Hypotype; dorsal view. (NZGS - TM 2635). Loc. S68/f257; GS277, Bobys Creek, Waipara, N Canterbury. Haumurian. [After Wilckens, 1922, pl. 3, f.3a, b].

Fig. 2. *Eriphyla meridiana* (Woods). (Bivalve). Hypotype; right valve. (NZGS - TM 2550). Loc. S56/f25; GS2, Haumuri Bluff, S Marlborough. Piripauan. [After Woods, 1917, pl. 15, f.2a, b].

Fig. 3. *Lucina canterburiensis* Woods. (Bivalve). Lectotype; right valve. (NZGS - TM 2551). Loc. S74/f17; GS23, Selwyn River, Malvern Hills, Canterbury. Haumurian. [After Woods, 1917, pl. 16, f.4].

Fig. 4. *Lahillia (Lahilleona)* cf. *neozelanica*. Finlay & Marwick. (Bivalve). Hypotype; left valve. (NZGS - TM 2568). Loc. S68/f263; GS761, Middle Waipara, Canterbury. Piripauan - Haumurian. [After Woods, 1917, pl. 16, f.8a, b].

Fig. 5. *Maorites tenuicostatus* (Marshall). (Ammonite). Hypotype. (AU - C 1032). Loc. N28/f626, Bulls Point, Kaipara Harbour, Northland. Piripauan - Haumurian. [After Henderson, 1970, pl. 9, f.3a, b].

Piripauan - Haumurian. [After Henderson, 1970, pl. 11, f.1a-c].

Fig. 6. *Vertebrites murdochi* Marshall. (Ammonite). Lectotype. (NZGS - CE 582). Loc. N11/f499; GS6959, Nedlers Point, Whangaroa Harbour, Northland. Piripauan - Haumurian. [After Marshall, 1926, pl. 30, f.1].

Fig. 7. *Kossmaticeras (Natalites) sulcatum* (Marshall). (Ammonite). Hypotype. (AU - C 1025). Loc. N28/f626, Bulls Point, Kaipara Harbour, Northland. Piripauan - Haumurian. [After Henderson, 1970, pl. 5, f.1a, b].

Fig. 8. *Gunnarites zelandicus* (Marshall). (Ammonite). Hypotype. (AU - C 1035). Loc. N28/f640, Whakapirau Creek, Kaipara Harbour, Northland.

Fig. 9. *Kossmaticeras (Natalites) bensoni* Henderson. (Ammonite). Holotype. (OU 4679). Loc. S163/f511, Fairfield Quarry, Fairfield, Dunedin District, Otago. Haumurian. [After Henderson, 1970, pl. 4, f.3].

Fig. 10, 11. *Baculites rectus* Marshall. (Ammonite). Hypotype; dorsal and lateral views. (VUW - C 210). Loc. N28/f636, Batley, Kaipara Harbour, Northland. Piripauan - Haumurian. [After Henderson, 1970, pl. 3, f.3a-c].

Plate 24 Paleocene molluscs from Eastern Otago.

Fig. 1. *Spineilo elongata* (Marshall). Holotype; left valve exterior. (OM). Loc. S180/f500, Mitchells Point, Wangaloa, SE Otago. Wangaloan (= Teurian?) (Paleocene). [After Marshall, 1917, pl. 36, f.41; Finlay & Marwick, 1937, pl. 1, f.4].

Fig. 2. *Lahillia (Lahilleona) neozelanica* Marshall & Murdoch. Hypotype, left valve interior. (NZGS - TM 4508). Loc. S180/f500; GS887, Mitchells Point, Wangaloa, SE Otago. Wangaloan (= Teurian?) (Paleocene). [After Finlay & Marwick, 1937, pl. 4, f.9, 10].

Fig. 3. *Cucullaea (Cucullastis) barbara* Finlay & Marwick. Paratype, right valve exterior. (NZGS - TM 4124). Loc. S163/f486; GS10195, Boulder Hill, Dunedin. Wangaloan (= Teurian?) (Paleocene). [After Finlay & Marwick, 1937, pl. 1, f.11, 14].

Fig. 4. *Heteroterma zelandica* Marshall. Hypotype. (NZGS - TM 5743). Loc. S163/f486; GS10195, Boulder Hill, Dunedin. Wangaloan (= Teurian?) (Paleocene). [After Finlay & Marwick, 1937, pl. 10, f.9, 10].

Fig. 5. *Zeacolpus (Leptocolpus) semiconcavus* (Suter). Hypotype. (NZGS - TM 4433). Loc. S163/f487; GS10194, Boulder Hill, Dunedin. Wangaloan (= Teurian?) (Paleocene). [After Finlay & Marwick, 1937, pl. 5, f.13; Marwick, 1971, pl. 7, f.3].

Fig. 6. *Kaitangata hendersoni* Finlay & Marwick. Holotype. (NZGS - TM 5744). Loc. S180/f500; GS887, Mitchells Point, Wangaloa, SE Otago. Wangaloan (= Teurian?) (Paleocene). [After Finlay & Marwick, 1937, pl. 5, f.4, 5].

Fig. 7. *Struthioptera osiris* Finlay & Marwick. Holotype. (NZGS - TM 5745). Loc. S180/f500; GS887, Mitchells Point, Wangaloa, SE Otago. Wangaloan (= Teurian?) (Paleocene). [After Finlay & Marwick, 1937, pl. 8, f.8-10].

Fig. 8. *Conchothyra australis* (Marshall). Hypotype. (NZGS - TM 5746). Loc. S180/f500; GS887, Mitchells Point, Wangaloa, SE Otago. Wangaloan (= Teurian?) (Paleocene). [After Finlay & Marwick, 1937, pl. 8, f.5, 6].

Fig. 9. *Ongleya tholispira* Finlay & Marwick. Holotype. (NZGS - TM 5747). Loc. S180/f500; GS887, Mitchells Point, Wangaloa, SE Otago. Wangaloan (= Teurian?) (Paleocene). [After Finlay & Marwick, 1937, pl. 13, f.8].

Fig. 10. *Campylacrum sanum* Finlay & Marwick. Holotype. (NZGS - TM 5749). Loc. S180/f500; GS887, Mitchells Point, Wangaloa, SE Otago. Wangaloan (= Teurian?) (Paleocene). [After Finlay & Marwick, 1937, pl. 12, f.4].

Plate 25 Representative Paleocene and Eocene macrofossils.

Fig. 1. *Serripecten venosus* (Hutton). Holotype; right valve exterior. (NZGS - TM 2842). Loc. Unknown, (?Bridge Point, Kakanui, Otago. Runangan (Up. Eocene)). [After Marwick, 1943, pl. 25, f.1; Boreham, 1965, pl. 5, f.5].

Fig. 2. *Cubitostrea gudexi* (Suter). Right valve exterior. (NZGS). Loc. N28/- (unregistered), Pahi Point, Kaipara Harbour, Auckland. Bortonian (Up. Eocene).

Fig. 3. *Venericardia (Fasciculicardia) acanthodes* (Suter). Left valve exterior. (NZGS). Loc. S127/f457; GS1988, Waihao Downs, S Canterbury. Bortonian (Up. Eocene).

Fig. 4. *Pseudoportlandia solenelloides* (Marshall). Right valve, interior and exterior. (NZGS). Loc. S146/f496; GS501, Hampden Beach, Otago. Bortonian (Up. Eocene).

Fig. 5. *Notostrea tarda* (Hutton). Left valve interior. (NZGS). Loc. CH/f3; GS1177, Tioriori, Chatham Islands. Waipawan (Up. Paleocene).

Fig. 6. *Notogenota finlayi* Powell. (NZGS). Loc. S127/f803; GS9508, McCullough's Bridge, S Canterbury. Kaiatan (Up. Eocene).

Fig. 7. *Carinacca waihaoensis* (Suter). (NZGS). Loc. S127/f803; GS9508, McCullough's Bridge, S Canterbury. Kaiatan (Up. Eocene).

Fig. 8. *Athleta necopinata necopinata* (Suter). (NZGS). Loc. S111/f506; GS5467, Lower Pareora Gorge, S Canterbury. Bortonian (Up. Eocene).

Fig. 9. *Monalaria concinna* (Suter). (NZGS). Loc. S127/f520; GS933, Black Point coal mine, Waitaki Valley, N Otago. Bortonian (Up. Eocene).

Fig. 10. *Portunites granulifer* Glaessner. (Crab). Holotype; dorsal view of carapace and right chela. (NZGS - DC 232). Loc. S172/f520; GS2977, Snowdrift Quarry, Milburn, S Otago. Bortonian (Up. Eocene). [After Glaessner, 1960, pl. 3, f.7].

Fig. 11. *Mauira curvispina* (Marwick). Holotype. (NZGS - TM 5751). Loc. S127/- (unregistered), greensand, Waihao Downs, S Canterbury. Bortonian (Up. Eocene). [After Marwick, 1926a, pl. 69, f.13].

Fig. 12. *Speightia spinosa* (Suter). Paratype. (NZGS - TM 5753). Loc. S127/f327; GS480, Waihao River, S Canterbury. Bortonian (Up. Eocene).

Fig. 13. *Fascioplex liraecostata* (Suter). Loc. S127/- (unregistered). Waihao Downs, S Canterbury. Bortonian (Up. Eocene).

Fig. 14. *Colubraria neozelanica* Maxwell. Holotype. (NZGS - TM 3875). Loc. S127/f803; GS9508, McCullough's Bridge, S Canterbury. Kaiatan (Up. Eocene). [After Maxwell, 1966, f.18, 20].

Plate 26 Representative Oligocene macrofossils.

Fig. 1. *Venericardia (Megacardita) ponderosa* (Suter). Left valve interior and exterior. (NZGS). Loc. S117/f492; GS1341, Wharekuri, Waitaki Valley, N Otago. Duntroonian (Mid. Oligocene).

Fig. 2. *Lima colorata paleata* Hutton. Lectotype; left valve exterior. (NZGS - TM 2915). Loc. S82/- (unregistered). Curiosity Shop, Rakaia River, Canterbury. Waitakian (Up. Oligocene). [After Suter, 1914, pl. 10, f.3; Boreham, 1965, pl. 11, f.6].

Fig. 3. *Serripecten (Janupecten) uttleyi* (Marwick). Holotype; right valve exterior. (NZGS - TM 4246). Loc. S117/f467; GS486, Wharekuri, Waitaki Valley, N Otago. Whaingaroan (Lr Oligocene). [After Marwick, 1924a, pl. 6, f.11].

Fig. 4. *Notocorbula humerosa* (Hutton). Left valve exterior. (NZGS). Loc. S127/- (unregistered). Otiake, Waitaki Valley, N Otago. Waitakian (Up. Oligocene).

Fig. 5. *Spissatella subobesa* (Marshall & Murdoch). Paratype; left valve interior and exterior. (NZGS - TM 4301). Loc. S127/- (unregistered). Wharekuri, Waitaki Valley, N Otago. Duntroonian (Mid. Oligocene). [After Marshall & Murdoch, 1919, pl. 19, f.3].

Fig. 6. *Cirsotrema caelicola* Finlay. (NZGS). Loc. S136/f544; GS3870, Campbell's Beach, All Day Bay, N Otago. Waitakian (?) (Up. Oligocene).

Fig. 7. *Magnatica planispira* (Suter). (NZGS). Loc. S127/f805; GS9517, Otekaike, Waitaki Valley, N Otago. Waitakian (Up. Oligocene).

Fig. 8. *Palaeeudyptes* sp. (Penguin). Hypotype; right femur, distal end. (NZGS - CD 33). Loc. N73/f806; GS9165, Motutarakatua Point, Kawhia Harbour, SW Auckland. Whaingaroan (Lr Oligocene). [After Marples & Fleming, 1963, f.1].

Fig. 9. *Histocidaris mckayi* Fell. Paratype. (NZGS - EC 137). Loc. S127/f376; GS482, Waihao Limestone, S Canterbury. Duntroonian (Mid. Oligocene). [After Fell, 1954, pl. 14, f.E].

Fig. 10. *Kekenodon onamata* Hector. (Archaeocete cetacean tooth). (NZGS - CD 40/1). Loc. S127/f379; GS476, Wharekuri, Waitaki Valley, N Otago. Duntroonian (Mid. Oligocene).

Plate 27 Representative Miocene bivalves.

Fig. 1. *Serripecten hutchinsoni* (Hutton). Hypotype; right valve exterior. (NZGS - TM 2841). Loc. S2/f520; GS5572, coast south of Anaweka River, NW Nelson. Altonian (Lr Miocene). [After Boreham, 1965, pl. 8, f.1].

Fig. 2. *Bassina speighti* (Suter). Hypotype; right valve exterior. (NZGS - TM 4542). Loc. N162/f499; GS4862, Wainuioru River, Wairarapa. Tongaporutuan (Up. Miocene). [After Fleming, 1962a, pl. 2].

Fig. 3. *Venericardia (Fasciculicardia) subintermedia* (Suter). Right valve exterior and part left valve interior. (NZGS). Loc. S167/f599; GS5095, Slip Point, Clifden, Southland. Clifdenian (Mid. Miocene).

Fig. 4. *Sectipecten wollastoni* (Finlay). Lectotype; right valve exterior. (NZGS - TM 2761). Loc. S51/- (unregistered). Callaghan's Creek, Westland. Kapitean (Up. Miocene). [After Boreham, 1965, pl. 7, f.3].

Fig. 5. *Lima colorata colorata* Hutton. Lectotype; left valve exterior. (NZGS - TM 2912). Loc. S136/- (unregistered). Awamoa Creek, Oamaru, N Otago. Altonian (Lr Miocene). [After Suter, 1914, pl. 10, f.5; Boreham, 1965, pl. 11, f.5, 7].

Fig. 6. *Lentipecten hochstetteri* (Zittel). Right valve exterior. (NZGS). Loc. S111/f526; GS5177, Tengawai River, S Canterbury. Otaian (Lr Miocene).

Fig. 7. *Cucullaea (Latiarca) hamptoni* Marwick. Holotype; left valve exterior and interior. (NZGS - TM 3774). Loc. N100/f31; GS1113, Rapanui River, N Taranaki. Tongaporutuan (Up. Miocene). [After Marwick, 1965, pl. 5, f.5; pl. 6, f.2].

Fig. 8. *Spissatella trailli* (Hutton). Lectotype; right valve exterior and interior. (NZGS - TM 2859). Loc. S136/- (unregistered). Awamoa Creek, Oamaru, N Otago. Altonian (Lr Miocene). [After Suter, 1914, pl. 13, f.4; Boreham, 1965, pl. 14, f.6, 8].

Plate 28 Representative Miocene gastropods.

Fig. 1. *Austrofusus (Austrofusus) coerulescens* (Finlay). Hypotype. (NZGS). Loc. S51/f583; GS4588, Goldsborough - Kumara Road, Westland. Kapitean (Up. Miocene). [After Grindley *et al.*, 1959, pl. 5, f.5].

Fig. 2. *Austrofusus (Neocola) apudalpha* Finlay. (NZGS). Loc. S111/f500; GS4795, Blue Cliffs, south Canterbury. Otaian (Lr Miocene).

Fig. 3. *Austrofusus (Neocola) alpha* Finlay. (NZGS). Loc. S167/f485; GS2948, Fouraye Siltstone, Clifden, Southland. Altonian (Lr Miocene).

Fig. 4. *Austrofusus (Neocola) beta* Finlay. (NZGS). Loc. S167/f483; GS2946, Calamity Point, Clifden, Southland. Altonian (Lr Miocene).

Fig. 5. *Austrofusus (Neocola) gamma* Finlay. (NZGS). Loc. S167/f477; GS2936, Slip Point, Clifden, Southland. Clifdenian (Mid. Miocene).

Fig. 6. *Austrofusus (Zelandiella) subnodosa* (Hutton). (NZGS). Loc. S110/f470; GS165, White Rock River, S Canterbury. Altonian (Lr Miocene).

Fig. 7. *Austrofusus (Austrofusus) magnificus* Finlay. (NZGS). Loc. S167/f477; GS2936, Slip Point, Clifden, Southland. Clifdenian (Mid. Miocene).

Fig. 8. *Bathytoma (Bathytoma) bartrumi* Laws. (NZGS). Loc. N28/f687; GS3245, Pakaurangi Point, Kaipara, Northland. Altonian (Lr Miocene).

Fig. 9. *Austrotoma obsoleta* Finlay. (NZGS). Loc. S68/f272; GS3209, Lower Waipara River, north Canterbury. Waiauan (Mid. Miocene).

Fig. 10. *Austrotoma echinata* Powell. (NZGS). Loc. S167/f477; GS2936, Slip Point, Clifden, Southland. Clifdenian (Mid. Miocene).

Fig. 11. *Maorivetia brevirostris* (Hutton). (NZGS). Loc. S111/- (unregistered). Sutherlands, Tengawai River, S Canterbury (*ex* Suter collection). Altonian (Lr Miocene).

Fig. 12. *Austrofusus (Zelandiella) fatua* Finlay. (NZGS). Loc. S68/f272; GS3209, Lower Waipara River, N Canterbury. Waiauan (Mid. Miocene).

Fig. 13. *Alcithoe (Mauithoe) insignis* Marwick. (NZGS). Loc. S68/f222; GS3840, mouth Dovedale Stream, N Canterbury. Waiauan (Mid. Miocene).

Fig. 14. *Alcithoe (Waihaoia) phymatias* (Finlay). (NZGS). Loc. S167/f477; GS2936, Slip Point, Clifden, Southland. Clifdenian (Mid. Miocene).

Fig. 15. *Amalda (Baryspira) stortha* (Olson). Holotype. (NZGS - TM 5754). Loc. S167/f479; GS2937, Long Beach, Clifden, Southland. Altonian (Lr Miocene). [After Olson, 1956, pl. 4, f.1, 2].

Fig. 16. *Opella subfimbriata* (Suter). (NZGS - TM 5038). Loc. N28/f687; GS1189, Pakaurangi Point, Kaipara, Northland. Otaian (Lr Miocene).

Fig. 17. *Polinices (Polinices) huttoni* von Ihering. (NZGS). Loc. S110/f470; GS165, White Rock River, S Canterbury. Altonian (Lr Miocene).

Fig. 18. *Conus hendersoni* Marwick. Holotype. (NZGS - TM 5755). Loc. N72/f489; GS1091, Waiapu Survey District, Gisborne. Kapitean (Up. Miocene). [After Marwick, 1931, pl. 15, f.273, 274].

Fig. 19. *Struthiolaria (Struthiolaria) subspinosa* Marwick. Holotype. (NZGS - TM 5758). Loc. S110/f470; GS165, White Rock River, S Canterbury. Altonian (Lr Miocene). [After Marwick, 1924b, pl. 11, f.4, 9].

Fig. 20. *Struthiolaria (Callusaria) spinosa* Hector. Neotype. (NZGS - TM 5760). Loc. S110/f470; GS165, White Rock River, S Canterbury. Altonian (Lr Miocene). [After Marwick, 1924b, pl. 11, f.12, 14].

Fig. 21. *Struthiolaria (Callusaria) callosa* Marwick. (NZGS). Loc. N165/f624; GS5740, Hurupi Stream, Palliser Bay, E Wellington. Tongaporutuan (Up. Miocene).

Plate 29 Representative Miocene molluscs with tropical affinities.

Fig. 1. *Chama huttoni* Hector. Holotype; left valve interior. (NZGS - TM 4289). Loc. N159/f446; GS81, Castlepoint, E Wellington. Nukumaruan (Lr Pleistocene). [After Suter, 1915, pl. 6, f.3].

Fig. 2, 3. *Spondylus aucklandicus* Marshall. Left valve exterior and interior. (NZGS). Loc. N28/f687; GS1189, Pakaurangi Point, Kaipara, Northland. Altonian (?) Lr Miocene.

Fig. 4. *Tapes parki* (Marshall & Murdoch). Hypotype; left valve exterior. (NZGS - TM 3903). Loc. S117/f678; GS9514, Aviemore Dam, Waitaki Valley, S Canterbury. Duntroonian (Mid. Oligocene). [After Maxwell, 1969, pl. 3, f.39, 40].

Fig. 5. *Conilithes wollastoni* Maxwell. (NZGS). Loc. S110/f470; GS165, White Rock River, S Canterbury. Altonian (Lr Miocene).

Fig. 6. *Cypraea (Notadusta) clifdenensis* Cernohorsky. (NZGS). Loc. S167/f598; GS5100, Long Beach Clifden, Southland. Altonian (Lr Miocene).

Fig. 7. *Chicoreus (Siratus) komiticus* (Suter). (NZGS). Loc. N28/f687; GS1161, Pakaurangi Point, Kaipara, Northland. Altonian (Lr Miocene).

Fig. 8. *"Marshallena" decens* Marwick. (NZGS). Loc. N105/f485; GS1579, Wahanui Road, Wairoa, Hawkes Bay. Opoitian (Lr Pliocene).

Fig. 9. *Ficus altispira* (Pritchard). (NZGS). Loc. S117/f495; GS1823, Wharekuri, Waitaki Valley, N Otago. Duntroonian (Mid. Oligocene).

Fig. 10. *Lyria zelandica* Finlay. (NZGS). Loc. S167/f598; GS10365, Long Beach, Clifden, Southland. Altonian (Lr Miocene).

Fig. 11. *Morum (Oniscidea) finlayi* (Laws). (NZGS). Loc. S167/- (unregistered). North bank, Clifden, Southland. Lillburnian (Mid. Miocene).

Fig. 12. *Echinophoria hectori* (Abbott). (NZGS). Loc. N28/f687; GS1189, Pakaurangi Point, Kaipara, Northland. Altonian (Lr Miocene).

Plate 30 Representative Pliocene molluscs.

Fig. 1. *Mesopeplum (Borehamia) crawfordi* (Hutton). Lectotype; right valve exterior. (NZGS - TM 2844). Loc. N158/- (unregistered), Taueru, Wairarapa. Waipipian (Mid. Pliocene). [After Boreham, 1965, pl. 10, f.1, 2].

Fig. 2. *Chlamys (Phialopecten) triphooki triphooki* (Zittel). Hypotype; right valve exterior. (NZGS - TM 2719). Loc. N134/f471; GS702, Napier, Hawkes Bay. Early Nukumaruan (Lr Pleistocene). [After Marwick, 1965, pl. 1, f.2].

Fig. 3. *Lima waipipiensis* Marshall & Murdoch. Holotype; left valve exterior. (NZGS - TM 4268). Loc. N137/- (unregistered). Waipipi Beach, Waverley. Waipipian (Mid. Pliocene). [After Marshall & Murdoch, 1919, pl. 19, f.1].

Fig. 4. *Trachycardium (Ovicardium) rossi* Marwick. Holotype; left valve exterior and interior. (NZGS - TM 4504). Loc. N106/f477; GS1541, Te Reinga Falls, Wairoa, Hawkes Bay. Waipipian (Mid. Pliocene). [After Marwick, 1944, pl. 35, f.7, 8].

Fig. 5. *Maorimactra chrydaea* (Suter). Right valve interior. (NZGS). Loc. S175/f594; GS5626, Te Waewae Bay, Southland. Opoitian (Lr Pliocene).

Fig. 6. *Hartungia typica typica* Bronn. (NZGS - TM 4671). Loc. N51/f518; GS5513, Kaawa Creek, SW Auckland. Opoitian (Lr Pliocene).

Fig. 7. *Clavatoma pulchra* Powell. Holotype. (NZGS - TM 5761). Loc. N115/f482; GS1568, Waihua River, Wairoa, Hawkes Bay. Waipipian (Mid. Pliocene). [After Powell, 1942, pl. 12, f.8].

Fig. 8. *Austrofusus (Austrofusus) pagoda* (Finlay). (NZGS). Loc. S51/f13; GS154, Kaniere, Westland. Waipipian (Mid. Pliocene).

Fig. 9. *Polinices (Polinices) waipipiensis* (Marwick). Holotype. (NZGS - TM 5762). Loc. N137/f303; GS1101, Waipipi Beach, Waverley. Waipipian (Mid. Pliocene). [After Marwick, 1924c, pl. 59, f.11].

Fig. 10. *Struthiolaria (Pelicaria) rugosa* Marwick. (NZGS). Loc. N150/f440; GS2855, Dannevirke, S Hawkes Bay. Early Nukumaruan (Lr Pleistocene).

Fig. 11. *Struthiolaria (Pelicaria) canaliculata* (Zittel). (NZGS). Loc. S29/- (unregistered), Awatere River, Marlborough. Waipipian (Mid. Pliocene).

Fig. 12. *Struthiolaria (Pelicaria) zelandiae* (Marshall & Murdoch). (NZGS). Loc. N137/f303; GS1101, Waipipi Beach, Waverley. Waipipian (Mid. Pliocene).

Plate 31 Representative Pleistocene molluscs.

Fig. 1. *Pecten benedictus marwicki* (Finlay). Hypotype; right valve exterior. (NZGS - TM 2030). Loc. N137/f512; GS4021, Castlecliff, Wanganui. Castlecliffian (Mid. Pleistocene). [After Fleming, 1957b, pl. 2, f.3].

Fig. 2. *Lutraria solida* (Hutton). Hypotype; right valve interior. (NZGS - TM 2903). Loc. N137/f348; GS1164, Nukumaru Beach, Wanganui. Nukumaruan (Lr Pleistocene). [After Boreham, 1965, pl. 19, f.2].

Fig. 3. *Barytellina tertia* Fleming. Hypotype; right valve interior. (NZGS - TM 4753). Loc. N42/f14; GS3528, Otahuhu well, Auckland. Waipipian (Mid. Pliocene). [After Marwick, 1948, pl. 2, f.10, 11].

Fig. 4. *Pecten novaezelandiae tainui* (Finlay). Hypotype; right valve exterior. (NZGS - TM 2016). Loc. N137/f501; GS4012, Castlecliff, Wanganui. Castlecliffian (Mid. Pleistocene). [After Fleming, 1951, pl. 1, f.4].

Fig. 5. *Protothaca (Tuangia)* aff. *crassicosta* (Deshayes). Right valve exterior. (NZGS). Loc. N117/f491; GS689, Mahia Peninsula, NE Hawkes Bay. Hawera Series (Up. Pleistocene).

Fig. 6. *Eumarcia plana* Marwick. Hypotype; right valve, interior and exterior. (NZGS - TM 4709). Loc. CH/f4; GS1178, Titirangi, Chatham Islands. Nukumaruan (Lr Pleistocene). [After Marwick, 1928, f.81].

Fig. 7. *Glycymeris (Glycymeris) shrimptoni* Marwick. Holotype; right valve exterior and interior. (NZGS - TM 4145). Loc. N134/f463; GS191, Kikowhero Stream, Hawkes Bay. Nukumaruan (Lr Pleistocene). [After Marwick, 1923, pl. 5, f.3; pl. 6, f.3].

Fig. 8. *Bassina parva parva* Marwick. Holotype; right valve exterior. (NZGS - TM 2207). Loc. N134/f482; GS1089, Okauawa Stream, Hawkes Bay. Nukumaruan (Lr Pleistocene). [After Marwick, 1927, pl. 47, f.153, 155].

Fig. 9. *Struthiolaria (Pelicaria) convexa* Marwick. Holotype. (NZGS - TM 5763). Loc. N134/f482; GS1089, Okauawa Stream, Hawkes Bay. Nukumaruan (Lr Pleistocene). [After Marwick, 1924b, pl. 15, f.3, 4].

Fig. 10. *Austrofusus (Austrofusus) glans glans* (Röding). (NZGS). Loc. N62/f505; GS5833, Te Piki, Cape Runaway, East Cape. "Castlecliffian" (early Hawera Series?) (Mid. Pleistocene).

Fig. 11. *Austrofusus (Austrofusus) glans chathamensis* Finlay. (NZGS). Loc. N135/f517; GS5310, Cape Kidnappers, Hawkes Bay. Castlecliffian (Mid. Pleistocene).

Fig. 12. *Leucotina ambigua* (Hutton). Hypotype. (NZGS - TM 5764). Loc. N137/- (unregistered). "Castlecliff", Wanganui. Castlecliffian (Mid. Pleistocene). [After Fleming, 1966, pl. 139, f.1651].

Fig. 13. *Pterotyphis (Prototyphis) zelandicus* (Hutton). Hypotype. (NZGS - TM 3117). Loc. N137/- (unregistered). "Castlecliff", Wanganui. Castlecliffian (Mid. Pleistocene). [After Fleming, 1962b, pl. 1, f.16].

Fig. 14. *Struthiolaria (Pelicaria) vermis* (Martyn). (NZGS). Loc. N137/- (unregistered). "Castlecliff", Wanganui. Castlecliffian (Mid. Pleistocene).

Fig. 15. *Struthiolaria (Pelicaria) acuminata* Marwick. Hypotype. (NZGS - TM 5765). Loc. N162/f762; GS6677, Maurioho Stream, Wairarapa. Early Nukumaruan (Lr Pleistocene). [After Grindley *et al.* 1959, pl. 5, f.7].

Fig. 16. *Struthiolaria (Struthiolaria) fraseri* Hutton. Hypotype. (NZGS - TM 5766). Loc. N134/f490; GS1098, Maraekakaho, Hawkes Bay. Nukumaruan (Lr Pleistocene). [After Marwick, 1924b, pl. 13, f.8, 9].

Plate 32 Representative Pleistocene warm- and cold-water fossils.

Fig. 1. *Isognomon zealandicus* (Suter). Left valve interior. (NZGS). Loc. N137/- (unregistered). Nukumaru Beach, Wanganui. Nukumaruan (Lr Pleistocene).

Fig. 2, 3. *Chlamys (Zygochlamys) delicatula* (Hutton). Hypotypes; right and left valve exteriors. (NZGS - TM 2836, 2837). Loc. N159/f454; GS1829, Castlepoint, Wairarapa. Nukumaruan (Lr Pleistocene). [After Boreham, 1965, pl. 5, f.8, 10].

Fig. 4. *Aulacomya ater maoriana* (Iredale). Hypotype; left valve exterior. (NZGS - TM 4179). Loc. S36/f504; GS6665, trawled off Cape Campbell, Cook Strait. Hawera Series (Up. Pleistocene). [After Boreham & Fleming, 1957, f.4, no.3].

Fig. 5. *Anadara trapezia* (Deshayes). Left and right valves, interior and exterior. (NZGS). Loc. N14/- (unregistered). Kohukohu, Hokianga Harbour, Northland. Hawera Series (Up. Pleistocene).

Fig. 6. *Philippia lutea* (Lamarck). (NZGS). Loc. N134/f627; GS10868, Mangatahi River, Hawkes Bay. Nukumaruan (Lr Pleistocene).

Fig. 7. *Capulus uncinatus* Hutton. (NZGS). Loc. N137/- (unregistered). Landguard Bluff, Wanganui. Late Castlecliffian (Mid. Pleistocene).

Fig. 8. *Pterotyphis (Prototyphis) angasi powelli* (Fleming). Holotype. (NZGS - TM 3115). Loc. N137/- (unregistered). Castlecliff,

Wanganui. Castlecliffian (Mid. Pleistocene). [After Fleming, 1962b, pl. 1, f.12].

Fig. 9. *Balanus (Megabalanus) tubulatus* (Withers). Holotype; (attached to exterior of *Chlamys delicatula*). (NZGS). Loc. N159/f446; GS81, Castlepoint, Wairarapa. Nukumaruan (Lr Pleistocene). [After Withers, 1924, pl. 7, f.1].

Fig. 10. *Semicassis pyrum* (Lamarck). (NZGS). Loc. N137/f399; GS4068, Castlecliff, Wanganui. Castlecliffian (Mid. Pleistocene).

Fig. 11. *Eunaticina cincta* (Hutton). (NZGS). Loc. N62/f505; GS5833, Te Piki, Cape Runaway, East Cape. "Castlecliffian" (early Hawera Series?).

Fig. 12. *Argobuccinum tumidum* (Dunker). (NZGS - TM 3992). Loc. S136/f710; GS5674, Rifle Butts Beach, S Oamaru, N Otago. Hawera Series (Up. Pleistocene).

Fig. 13. *Calliostoma spectabile couperi* (Vella). Holotype. (NZGS - TM 4995). Loc. N165/f515; Makara River, Wairarapa. Early Nukumaruan (Lr Pleistocene). [After Vella, 1954, pl. 25, f.6].

Index of Fossil species by Genus